Web Sites for Culture

Essential Principles for Great Arts Web Sites

Eugene Carr

PATRON PUBLISHING
NEW YORK

Patron Publishing
A division of Patron Technology, Inc.
850 Seventh Avenue, Suite 801
New York, NY 10019

Cover design by Jonathan Gullery

Text layout by Michelle Kettner

Library of Congress Control Number: 2005906392

ISBN: 0-9729141-2-9

First Printing August 2005

Printed in the United States of America

About the Author

Eugene Carr is the founder and president of Patron Technology®, the leading online marketing software company for the arts & culture industry. PatronMail®, the company's e-mail marketing software system, is used by over 500 arts and cultural organizations in the U.S. and the U.K., including the New York Philharmonic, Pittsburgh Opera, Roundabout Theater and the Alvin Ailey Dance Company. The company also builds Web sites for arts organizations, and recent clients include the Washington National Opera, the Maine Center for the Arts, and Chanticleer.

Mr. Carr is a regular guest speaker at arts industry conferences and is the author of *Wired for Culture: How E-mail is Revolutionizing Arts Marketing* (2003) and *Sign Up for Culture: The Arts Marketer's Guide to Building an Effective E-mail List* (2004).

Prior to founding Patron Technology, Mr. Carr was the founder and president of CultureFinder.com, an award-winning arts information portal, which was funded by America Online and Comcast.

Acknowledgements

My thanks go to the members of my staff at Patron Technology and several friends who helped me hone my ideas and bring this book to reality.

Michelle Paul was the project manager; she also contributed sections of text and editorial changes. She deserves high praise.

Eric Zakim reprised the role he played for my last book by providing brilliant comments and editorial insight. Michelle Kettner did the layout and graphics with her usual speed and professionalism.

Further thanks go to Jane Chafin, JD Hixson, Lily Traub, Mark Famiglietti, and Kathleen Drohan for their editorial suggestions and endless re-readings.

Dedication

All too often arts managers are told they should run their organizations "more like a business." In fact, many arts institutions already operate in a more creative, clever, and resourceful way than the role models they are to emulate. But, when it comes to their Web sites, arts managers still have a lot to learn from corporate America.

This book is dedicated to the idea that arts Web sites will be so great that corporate managers across America will regularly think that they should run their Web sites "more like an arts organization."

Contents

CHAPTER 1

The Web Site Opportunity

Most arts Web sites I visit are marketing opportunities waiting to happen. They are good, but not great. I'm convinced that the arts marketers running these sites are in need of a few guiding principles in order to dramatically improve their results.

In this book, I offer a way of thinking about how to build effective arts Web sites. It is a companion to two others I've written about online marketing for the arts: *Wired for Culture: How E-mail is Revolutionizing Arts Marketing* and *Sign-Up for Culture: The Arts Marketer's Guide to Building an Effective E-mail List.*

My main message in those books is that e-mail marketing is the most effective tool for marketing the arts online. That's because e-mail notices sent to a large home-grown list can generate significant traffic to your Web site. So, e-mail is half of the online marketing equation for arts marketers.

However, once you have successfully motivated visitors to click over to your site from an e-mail, you need to provide them with an outstanding Web experience.

The graphic below illustrates how e-mail and Web sites work together.

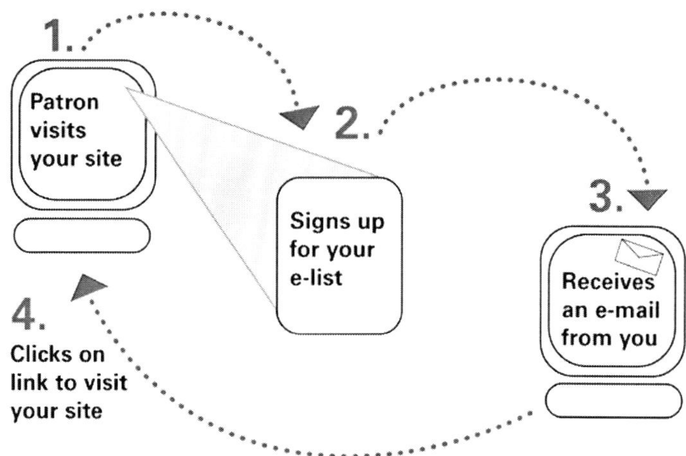

Thus, a great Web site is the other half of the online marketing equation, and the subject of this book.

The good news is that having a great arts Web site is not about having a big budget. And it is not about finding a great site designer or having the proverbial young person who "knows computers."

Rather, the key to building a great arts Web site is knowing some basic marketing and design principles, which will guide your thinking and help your decision-making process. My goal here is to make this book a resource to help ensure your site becomes a truly outstanding marketing tool. I've picked out only the key principles that will help you achieve world-class results, even within a typically constrained arts budget!

Before we get started, let me tell you what to expect from this book. First, it is intentionally basic. Success in marketing (like many other things in life) is about getting the basics right. Whether it's a golf swing or great diction on stage, most often greatness is borne from excellent fundamentals. Unfortunately, the basics frequently get left behind in a blaze of newfangled approaches, especially on the Web!

Second, this book is not a technical guide for Web programmers. It focuses on important issues of site strategy, design concepts, and traffic development, all written from a marketer's perspective.

If you are responsible for running a Web site, this book will help you make it more successful. If you're a technical site manager, you'll learn about Web sites from a marketing angle, which will help you communicate and strategize with your marketing colleagues. And if you are in charge of marketing for a very small institution or you are an individual working with an outside site designer, my aim is to give you the tools you'll need to get the best results from your vendor.

Finally, I expect this book may come in very handy at Board meetings when everyone on the Board seems to claim his or her own expertise about Web sites!

One note of importance: This book does not cover the vast and often complex subject of e-commerce software. If you're in the process of selecting a new ticketing or online fundraising system, a good place to start is by looking at two Web sites that cover these topics well: www.intix.org and www.ticketing.org.uk.

Now, let's get to work.

CHAPTER 2

Understanding the Problem

Why are most arts Web sites so lackluster?

When I claim that most arts Web sites could be better, I don't mean that I dislike their design or that they don't utilize the most current technology. What frustrates me is that these sites clearly don't support the marketing and fundraising goals of their organizations. Most are rather loosely organized, rambling, and ultimately ineffective.

I'm convinced that the root of this problem lies in the early years of the Web, when misinformation and misperceptions abounded as to what the Web really could offer marketers. The result is a generation of ineffective arts Web sites.

When consumer Web sites were first coming into public consciousness in the mid-1990s, the press promoted the notion that every site would immediately attract thousands of visitors. This amazing promise fueled the entire dot-com stock market bubble. It was a "something for nothing" scenario that had a particularly attractive allure for both marketers and investors.

News articles reported Web sites with zillions of "hits" (a truly useless term for understanding anything about consumer Web behavior). The press reported online audiences instantly arriving in hordes at new sites.

So, arts marketers quickly built Web sites, which were often nothing more than season brochures on the Web. Designed to offer a "salad bar" of information, most were casually organized and based on the idea that visitors would take the time to poke around and find the information that they were seeking.

Then serious e-commerce started to become a reality in the late 1990s. Around it, a sophisticated science of studying online consumer behavior developed. What we have leaned is that well-designed Web sites both enable consumers to get the information they want quickly, and effectively support the marketing objectives of an organization.

Web sites shouldn't just be an exhaustive collection of pages about an institution. Rather, they are marketing tools. As such, they need to be organized strategically, just like marketing tools in the off-line world.

Unfortunately, most arts marketers have not yet embraced this way of thinking, which accounts for the current state of affairs.

Web sites as "machines for marketing"

So, how do you produce a great site? Let's start with a definition:

> **A great arts Web site is a carefully structured collection of Web pages designed and organized to achieve a measurable goal, one that supports your organization's overall mission as well as your bottom line. A great arts Web site offers visitors the information they want in the fastest way, with the fewest number of mouse clicks and the least amount of confusion.**

Such a Web site is essentially a "machine for marketing." Just like a physical machine that has input and output, the input on a Web site is your patrons' time and attention, and the output is information that motivates them to think differently, participate in your programs, and—oftentimes—purchase services or tickets from you.

I encourage you to begin to think of your site in these terms: *What did the people who visited my site today come away with? More importantly, what were we left with as a result of their visit?*

What does "great Web design" mean anyway?

When I hear some arts marketers talk about great Web design, all too often what they mean is that the site is aesthetically pleasing to them.

The subjective notion that a site looks good, like a comment about a new suit or pair of shoes, is not really the ultimate goal. Designing a Web site in order to look good usually misses the point.

Of course, you should want to design something beautiful. But you also want it to be highly functional.

Think for a moment about your direct mail brochure. You design it to get the best return on your investment. Ultimately, you are judged by your return rate, and ROI of the effort overall. Whether your brochure has the most "cutting edge" or "cool" design is part of the equation, but not the whole story.

As you can see, Web site design is really a combination of art and science. It's more like designing a car, or a handbag, or even a high-quality ski jacket. In each of these cases you're dealing with a highly functional item that needs to meet a specific set of requirements for the user, and be visually and aesthetically pleasing to boot.

So, start thinking about your Web site as a very functional marketing tool within your arsenal of marketing techniques. You should build it (and judge it) with the same degree of care you use in assessing all your other marketing programs.

Summary Points:

★ Years of misinformation and misperception have created a generation of ineffective arts Web sites.

★ Web sites should be highly effective "machines for marketing."

★ Your goal should be to design something that is not only aesthetically pleasing but also highly functional.

CHAPTER 3

Definitions That Really Matter

Since I've promised that this won't be a technical book, I'll demystify Web technology in this chapter by offering definitions of a few important technical terms.

You may recognize some of these terms as ones bandied about by techies who don't really want you to understand what they are doing. So, here are plain-English definitions to help you understand what's really going on and talk with the people who will be programming your Web site.

Above the fold: The section of your e-mail or Web site that is visible to viewers when they first see your information on their screen, without their having to scroll down the page. "Above the fold" is related to the same newspaper term that refers to the top half of the front page of a printed newspaper, which is literally "above the fold" and thus the most visible.

Content: The text, images, and other media (such as audio or video) that appear on your site.

Domain name: A Web site address that has been registered to you (e.g., www.websitename.org).

Flash: A software language developed by a leading software company, Macromedia, which provides programmers with an expanded ability to display moving pictures and text on a Web page. According to Macromedia, Flash is installed on more than 97% of internet-enabled desktops and on many other popular electronic devices.

HTML: Hypertext Markup Language, or the basic coding language used for most Web pages and graphic-filled e-mails.

ISP: Internet Service Provider, or the company that provides internet access (such as AOL, Verizon, etc.). These companies also typically provide their customers an e-mail address or software to access their e-mail.

Navigation: The set of links that helps your visitors find information on your site. Typically, navigation links are found at the top or on the side of a Web page.

Operating system: Software that has been loaded on a computer to control the operations of that computer and provide a common platform to run other software programs. Currently, the most popular operating systems are Windows XP, Windows 98, and Mac OS X.

Pop-up pages: Content that appears in a separate window from the Web page—it "pops up" above the page you're viewing. This window can be closed without the user closing the main page.

Server: The physical computer that stores the content and data of your site. Your server is typically housed at a hosting company's location.

Splash page: An initial welcome page that typically contains nothing more than a logo and a link that says "click here to enter."

Streaming media: Audio or video content that plays as it downloads from the server instead of downloading in its entirety before playing.

URL: Uniform Resource Locator, or the Web address of a Web site (as in, "Here's my URL; it's www.laboratorytheatre.org").

Web host: A company that provides space on its computers to store the content of your site. These companies also supply the connectivity that allows your site to appear on the Web to the public.

In addition to technical definitions, you will need to know some terms associated with Web site statistics in order to judge whether your site is effective. Web site statistics are based on data generated by the server that operates your site. Data reported by the server will help you understand the activity on your site in a marketing context. Below are some key terms used in discussing these reports:

Traffic: An overall term that refers to how many visitors come to your site and what their activity is on your site.

Visits: The number of individual times your site gets viewed during a given period of time.

Unique visitors: The number of unique individuals who visit your site in a given time frame, as opposed to the total number of visits. For example, 200 visits could be generated by 100 unique visitors, each coming twice in a given month.

Page views: The number of times a page on your site is viewed by a visitor. For example, one "visitor" might look at three pages on the site, which would generate three page views.

Visit length: The amount of time each visitor spends looking at your site during a given session.

Hits: The number of times each individual file (text, graphics, etc.) downloads when a visitor views a Web page. It is not a Web marketing term, but a measure of how much work the server is doing to create the page, and thus only a crude measure of the number of visitors to a site.

CHAPTER 4

How Arts Patrons Behave Online

Many people have written about why the Web has so quickly become an integral part of our lives, and the explanations are varied and often complex. I see it in simpler terms. We're living in a world in which people are increasingly impatient. They value convenience more than almost anything else.

I think the growth in popularity of the Web can be succinctly explained in this way: the Web provides information to people **more quickly and easily than they could find it in any other way.**

If you think about it, the vast amount of information that we look for online every day is already available offline - it just takes longer to get. Calling for an airline schedule by phone or waiting for a box office person to tell you if there are seats available will get you the same result as clicking on a Web site, but it's more tedious.

So, what motivates Web usage is, first and foremost, **convenience.**

We can prove this by looking at some basic consumer behavior statistics. According to Nielsen/Netratings, during the month of June 2005 the average home Web user spent 28 hours online; they logged in about once a day and looked at an average of about 59 sites during that month, or an average of 1.79 sites per visit. The average Web session was 50 minutes, and the average time spent at any given Web page was 49 seconds.

This last statistic often shocks marketers who somehow imagine visitors spending hours pouring over their sites. In reality, very few Web users visit a Web site for longer than a few minutes.

Arts sites receive short visits too

In reviewing Web traffic reports for arts sites, similar patterns emerge.

The traffic numbers for the Kansas City Symphony from February 2005 illustrate this well:

Kansas City Symphony Web Site Traffic
(February 2005)

Total number of page views:	53,540
Unique number of visitors:	7,760
Total number of visits:	10,345
Average page views per visit:	5.31
Average time per visit:	2:58

Over the last decade of analyzing traffic patterns on arts Web sites, I'm continually amazed that, notwithstanding the often deep and rich content provided on most arts sites, typically it's the basic information related to schedule, calendar, location, and ticketing that comprises the most frequently viewed pages.

Simply put, arts patrons visit arts sites to find out what's going on, and how they can attend. Although this seems rather logical, I've found that often senior management or the marketing committee of the Board doesn't necessarily think that this is what the main focus should be.

So, let's continue to analyze the Kansas City Symphony's site, which will demonstrate this phenomenon clearly. On its main screen, there are six main sections.

Looking more closely, you'll see that each of these sections has several sub-sections. In total, there are about 28 sections in this site, comprising approximately 66 total pages.

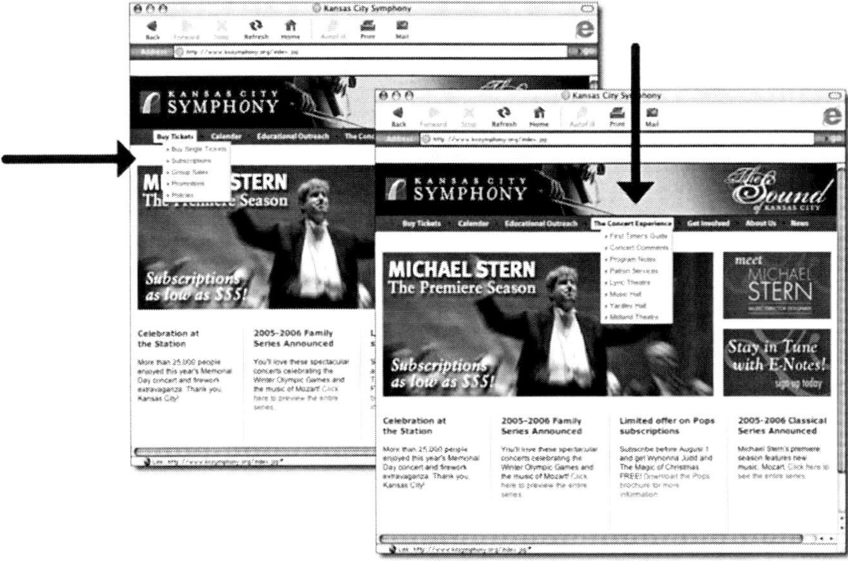

Now, here's a review of the site's top 10 pages viewed in February 2005.

Kansas City Symphony Web Site
Top 10 Pages Viewed
(February 2005)

1. Calendar of Events:	9,934
2. Concert Event Details:	8,444
3. Home Page:	8,160
4. Buy Tickets Transaction Window:	2,680
5. Concert Music Files:	2,150
6. Buy Single Tickets:	1,908
7. About Us – Job Openings:	1,190
8. Buy Tickets Shopping Cart:	972
9. The Symphony Experience – Music Hall:	958
10. Midland Theatre Seating Chart:	915

As you can see, of the 53,000 page views, 70% were related directly to the event-selection and ticket-buying experience. In other words, more than two thirds of the page views came from just 15% of its pages.

Let's look at another example, the Stamford Center for the Arts in Stamford, Connecticut.

Below is the main screen of the Stamford Center's site. There are five main navigation sections on this site with 31 sub-sections and 60 pages in total:

During the week of May 8-14, 2005, its Web site statistics looked like this:

Stamford Center for the Arts Web Site Traffic
(Week of May 8-14, 2005)

Total page views:	16,812
Total number of visits:	3,642
Pages viewed per visitor:	4.62
Average time per visit:	5:02

The list below summarizes the top 9 pages visited on this site during that week:

Stamford Center for the Arts Web Site
Top 9 Pages Viewed
(Week of May 8-14, 2005)

1. Main Screen	11,719
2. Calendar of Events	1,599
3. Event Details	1,456
4. General Information	216
5. Contact Us	160
6. Search Our Events	129
7. Directions and Parking	128
8. Renting Our Theatres	118
9. Seating Charts	111

About 93% of the total pages viewed relate to information about what's going on at the Center or information on contacting the Center to visit or rent it.

If you are not yet familiar with your own Web site traffic reports, I urge you to review them now. In all likelihood, you will find similar patterns. If you do not have access to your Web statistics, ask your site hosting company to provide them to you.

Summary Points:

★ The Web provides information faster and more easily than any other source.

★ On average, visitors spend only a few minutes on any particular arts Web site.

★ Visitors want to find basic information as quickly as possible, and with the fewest number of mouse clicks.

CHAPTER 5

Understanding Your Needs

Understand the needs of your Web audience

Now that you've learned about the overall behavior of online users, next you ought to know the particular needs of your site's visitors. How much do you really understand about their online interests?

Let's start with some basic questions to ask about your online audience:

- What are their expectations when they come to your site?
- What do they think they will be able to accomplish on your site (e.g., register for a class, buy a ticket, make a donation)?
- Have they heard about your organization elsewhere? And if so, what impressions do they bring with them? Are you supporting an existing image or creating a new one?
- What kind of information are they seeking, exactly?
- What portion of your visitors are:
    - Current ticket buyers, members, or fans?
    - First-time visitors for whom your site will be the first impression of your organization?
- How did they find your site? What motivated their visit?
- What is their impression of your site after viewing it for the first time?
- For regular visitors, what information have they returned to obtain? How frequently do they visit your site?
- Are there places on your site where they can easily make a purchase or transaction, ask a question or send an inquiry? To what degree does this capability matter to your organization?

Web audience research (without breaking the bank)

If you're not able to answer these questions with clarity and confidence, then you'll need to do some research. Though researching your site audience is essential, it need not be expensive or difficult.

In a perfect world with sufficient resources, you might engage a professional marketing research company for tens of thousands of dollars. They would use a combination of phone and Web-based tools to survey a statistically valid and pre-qualified audience.

Since I'm imagining that your organization cannot afford that approach, I want to offer a simple and cost-effective method that will still provide you with highly useful results.

I've used this technique successfully for years. It won't break the bank, and yet it gets you nearly instant results. Best of all, you can do it completely on your own with little if any technical assistance.

With this approach, you'll create a Web-based survey that you will prominently display on your site's main screen, collecting responses from those actually visiting your site. Depending on your site traffic, you'll leave the survey up for a few days or even a week or so.

There are many fine Web-based survey tools which simplify this process. The one we have used and recommend is SurveyMonkey (www.surveymonkey.com). With SurveyMonkey, you build and format your survey online, and it's simple to link to it from your site.

The software automates most of the process. It collects the responses, calculates and charts the results, and offers online tools for more sophisticated analyses. SurveyMonkey costs about $20 a month, and there are plenty of other similar resources to choose from. A quick search on Google.com for "Web-based survey software" will get you started.

As for the format of the questions, here is a survey for the fictional Laboratory Museum, based on one I prepared for a leading

American museum some years ago. It should provide a good starting point for your own work.

Laboratory Museum Web Site Survey

1. How frequently do you visit the Laboratory Museum's Web site ?
- This is my first visit
- A few times a year
- About once a month
- About once a week
- About twice a week
- Daily

2. If you had to select the one reason why you came to the site today, what would it be?
- Find out museum hours or location
- Find out admission prices or purchase tickets
- See a calendar of programs and events
- Get café and dining information
- Shop or browse museum store
- Preview or learn about our exhibitions
- See pictures of art
- Read information about art
- Find activities for kids or teens
- Learn about our history or mission
- Find out about membership or charitable giving

3. Please rank the importance of the following features on the Laboratory Museum's site.
- Find out museum hours or location
- Find out admission prices or purchase tickets
- Get café and dining information
- Shop or browse museum store
- See a calendar of programs and events
- Preview or learn about our exhibitions
- See pictures of art

- Read information about art
- Find activities for kids or teens
- Learn about our history or mission
- Find out about membership or charitable giving

4. What is your overall impression of how easy it is to use and navigate the Laboratory Museum's site?
- Difficult
- Somewhat difficult
- Neither difficult nor easy
- Somewhat easy
- Very easy

5. To what degree did you find the content of our site interesting and useful?
- Extremely
- Very
- Somewhat
- A little
- Not at all

6. Where do you live?
- New York City
- Other New York
- Connecticut
- New Jersey
- Other U.S.
- Other country (please specify)

7. Are you a member of the Laboratory Museum?
- Yes
- No

8. How much has your visit to the Laboratory Museum's Web site influenced your decision to visit ?
- Not at all
- Somewhat
- Very much

9. In total, approximately how many professional cultural arts events or exhibitions did you attend over the past 12 months? (Classical music, dance or ballet, Broadway production, theater, opera, art museum or exhibition, etc.)
- None
- 1-2
- 3-4
- 5-6
- 7-10
- 11-15
- 16-26
- 27 or more

10. How many times have you been to the Laboratory Museum in the past 12 months? (If you cannot recall exactly, please give us your best estimate.)
- None
- 1-2
- 3-4
- 5-6
- 7-10
- 11-15
- 16-26
- 27 or more

11. Please tell us your gender.
- Male
- Female

12. Please tell us your age.
(If under 18, skip to question 15.)
- under 18
- 18 – 24
- 25 – 34
- 35 – 44
- 45 – 54
- 55 – 64
- 65+

13. What is the highest level of education you have completed?
- Some High School
- Graduated High School
- Some College – no degree
- Associate's Degree
- Bachelor's Degree
- Post-Graduate Degree

14. What is your annual household income?
- Less than $20,000
- $20,000 - $39,999
- $40,000 - $49,999
- $50,000 - $74,999
- $75,000 - $99,999
- $100,000 - $149,999
- $150,000 - $199,999
- $200,000 - $249,999
- $250,000+
- Prefer not to answer

15. Is there anything else you would like us to know?

I encourage you to include this final question, "Is there anything else you would like us to know?" as a text box that allows for written comments.

Arts patrons tend to be rather motivated, so don't be surprised at the degree to which some will take the time to write lengthy commentary about their interests in your organization, or complaints about some aspect of your operation.

As indicated before, once you've built your survey, put a prominent link to it on your main screen.

Here's a good example:

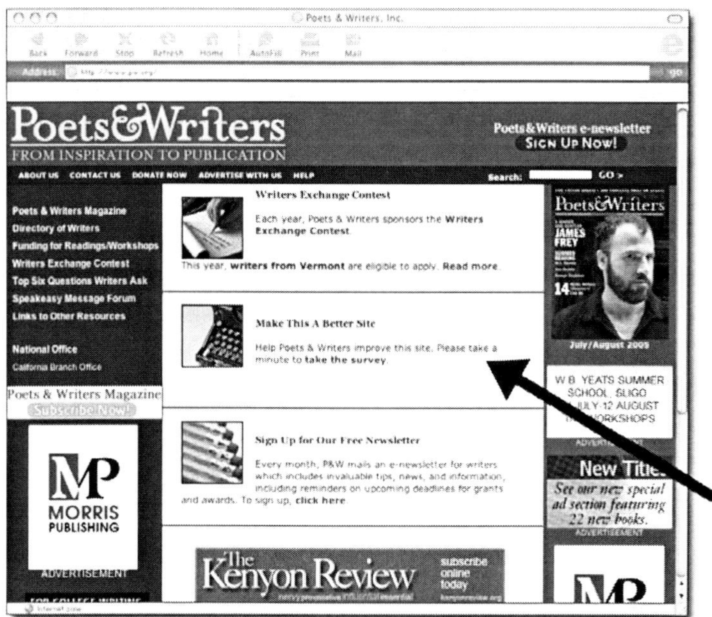

An even more effective approach (which will involve your site designer) is to make arrangements to have the survey pop up when visitors arrive at your site. If you do this, include some introductory text above the first question which says something like "Welcome! We want to know more about you and your needs..."

Of course, there will be those with pop-up blockers installed on their computers who will miss the survey, but this is worth doing nonetheless. We have never encountered an instance when people were annoyed at this approach. Most are very willing to give you their opinions!

Keep the link active until you've obtained at least 500 responses, and then remove it from your site. Now that you've completed the data collection process, your survey software will provide the initial results for your review. In addition, you're likely to have hundreds of comments to read though.

The analysis of this data should be fairly straightforward. In our experience, you'll either learn something new about your audience's online interests and needs, or at least have hard evidence to confirm your expectations.

Know your online competition

Since the arts patrons who visit your site also visit other arts organizations' Web sites, it's a good idea to visit these same sites to get to know your online competition.

I have found that one of the best ways to get new ideas for your Web site is to review and analyze what others in your industry are doing. This review isn't a competitive review about who is doing what next season, but rather a Web site functionality and design review.

Do this analysis systematically, and take good notes. Here is a list of the types of sites you should review:

• ***Sites in your community:*** First, look carefully at your direct competitors and at other organizations in your community.

Look at organizations of all sizes. I find that it's often the smaller organizations that manage to "get it right" or offer something unique or special that might inspire your thinking.

Don't limit yourself to your genre. Your patrons go to a wide variety of other arts events, so even if you don't think you compete with the symphony or the local university arts series, review those sites anyway.

• ***Outside your community:*** Since the Web gives you such easy access to sites all over the world, pick 20 sites in your genre to review.

You'll be amazed at what you can learn, particularly when you take a good look at sites abroad.

• ***"Big-name" large budget organizations:*** Pick 20 of the largest and most famous organizations you can think of, ones with the largest budgets.

It's not often that I see a direct correlation between big budgets and excellent sites. Often large sites are too busy and filled with too much useless technology.

Nonetheless, there are often hidden gems to inspire you. See for yourself how good a job they have done, and borrow or adapt as many good ideas as you can. Most importantly, pay attention to where you feel they fall short.

Create a functionality list

As you do your research, keep notes in two ways. First, keep a list of the sites you particularly like and what it is about them that you find particularly good.

Conversely, record a list of sites you think are terrible, with the same annotations. That list will become very useful if you are working with a designer or an outside design firm. Offering specific examples to your design team about what you like and don't like will be an invaluable aid. (More about that in Chapter 11.)

Second, make a list of the functionality offered by the sites you're comparing and assess whether that functionality is needed on your site. I find that making a chart similar to the one below is very helpful.

Web Site Functionality Chart

FUNCTION:	Online Donations
Existing on our site now?	No
Needed on new site?	Yes
Example site:	www.laboratorytheatre.org
Comments:	Confirmation screen design clean and simple

FUNCTION:	Address and parking information
Existing on our site now?	Yes
Needed on new site?	Yes
Example site:	www.denverartmuseum.org
Comments:	Make sure to include parking garage rates

FUNCTION:	Information for teachers
Existing on our site now?	No
Needed on new site?	Yes
Example site:	www.gevatheatre.org
Comments:	What teachers need to know about bringing their students to our performances

Summary Points:

★ You need to understand the particular needs of your site's visitors.

★ Research surveys can be done easily and inexpensively.

★ Review and analyze what others in your industry are doing with their Web sites.

★ Based on your research, create a functionality list of what is needed on your Web site.

CHAPTER 6

Developing a Web Site Strategy

You must have a clearly articulated goal

If you take away only one principle from this book it should be this: **To have an effective site, you must have a central goal.**

I'd like you to try to answer the following question in one sentence:

What do you want your Web site to achieve?

It sounds rather simplistic, but I'm amazed how many arts marketers can't answer this very question when I pose it to them.

It always reminds me of the saying, "If you don't know where you're going, you'll never get there." Quite simply, without a clear definition of your marketing objective, your site will be nothing more than a missed marketing opportunity.

It's foolhardy to think that your Web site can be all things to all people. Yes, you can have many overlapping objectives. But from a marketer's perspective, when you focus on the one goal that is most important for your Web site, you'll force yourself to focus your Web site's design and performance to acheive it.

So, the recipe for developing your site's goal must take into account two basic factors: what your patrons want your site to provide, and what your organization's overall marketing goal is. In most cases, these two areas will coincide nicely. If not, you'll have to work on developing a goal that effectively encompasses both of these needs. Only then will you be ready to start the process of designing your site.

Whether your goal is to bring in ticket sales, disseminate information to the public about your mission, or get people to visit a gallery or a show opening, your Web site should achieve a goal that you determine.

A clearly articulated goal will help you make decisions

The clearer you are about what your goal is, the easier the job of designing your site will be. Since we know how little time Web users actually spend on each site they visit, and how few pages they look at, your overarching task will be to translate your goal into a site design that increases the likelihood of a patron's visit to your site producing the results you want.

I often ask my clients to try to answer this question: ***If you could magically guide each site visitor to click on only one link on your site's main screen, what would that link be?*** If you truly have a well-articulated goal, it should be easy to answer this question.

For instance, people often ask me what the goal of my Web site would be were I still running the American Symphony Orchestra. Even though I've been out of the orchestra business for a decade, the answer is clear to me, and I think it applies to many other organizations.

My goal would be to collect as many e-mail addresses as possible. Therefore, if I got only one click, I'd want it to be on the link called "Get our e-mail newsletter."

I see it this way. If I can get someone to join my e-mail list, I can market to them forever. But if they come and go without leaving any trace, I've missed a huge online marketing opportunity to contact them directly.

With my goal firmly in mind, I can now approach the overall design of my site. When I review possible main screen designs, my reactions to different alternatives will not be based on whether I like a particular design because of its aesthetics, but rather: ***Is this design more likely to yield me an e-mail sign-up?***

Your site's central goal should be specific and measurable

When I talk with potential clients about their interest in overhauling or rebuilding their site, I pay a lot of attention to their answer to my very first question, which is always: "Why are you re-doing your site?"

If you are upgrading or rebuilding your site, take a moment and answer this question for yourself: ***Why are you re-doing your site? What do you wish to achieve that you are currently not achieving?***

Here are some typical answers:

> "We think it needs a new, fresh look."
>
> "Nobody can find any information on it."
>
> "It's not working for us."
>
> "It was designed by someone who left a long time ago, and we want to be able to update it ourselves."
>
> "It doesn't represent us well."

Sound familiar?

While each of these is a good and valid answer, what bothers me is that they express only a vague understanding of a problem and don't really suggest any solutions.

Let me share some answers I'd prefer to hear:

> "We want to sell 25% of all our tickets online because many of our colleagues in our community achieve this rate. But our site now generates only about 9%."
>
> "We know that the bulk of our visitors want show information, but right now it's buried on our site. We now have only 15% of our page views relating to the current show, and we want to raise that number to 50%."

> "We know that 70% of the people coming to our museum's site want to know what the special exhibition is. So, we want to have at least 25% of our entire site traffic view the 'current exhibition page.'"

Each of these answers demonstrates a very specific objective, along with a measurable goal that can be used as a benchmark for your design team and for later site analysis.

Take a moment now and focus your thinking on developing a revised, more specific answer to my first question: What it is that you want your site to accomplish?

Get agreement for your goal from your boss and your Board

Once you've settled on an objective, I recommend that you "sell it" to the rest of your staff and management. You might even want to have your Board vote on it.

Once you have organization-wide agreement on your objective, you will have a much easier task both developing your site and updating it over time.

Inevitably, there will be changes at your organization, events or circumstances that may crop up which will cause you to want to "throw out" all your hard strategic planning and simply put new information on your site's home page.

The adoption of specific, agreed-upon site goals will help guard against this from happening. I'd like to think that every time you work on your site, you'll repeat the mantra, "What is the ultimate goal of my site?"

Summary Points:

★ To have an effective site, you must have a well thought-out and researched goal that meets the needs of your visitors and your organization.

★ Work with your staff and management to adopt this goal as an organization-wide objective.

CHAPTER 7

Site Design Fundamentals

Your site's design influences the behavior of those visiting it

As you begin thinking about your site's design, the retail adage "location, location, location" is as true online as it is for brick-and-mortar shops. How and where you place links, images, and content on your site will affect how your visitors use your site.

The simple reason for this, as we have learned before, is that your visitors are busy and anxious to get on and off your site as fast as possible. So, they will see and react to what you put in front of them first. By the same token, what's buried deep within your site is less likely to be seen.

Years ago I worked with AOL, where we lived by the motto, "Every time you ask your visitors to click, 50% of them will leave instead." The more you make people click, the more likely they are to give up, get frustrated, and abandon your site.

What I find helpful as I work on arts sites is to put together a mental picture of an arts patron visiting the site.

My example is Sabrina. She is 41 years old and lives in Boston. She's someone who logs on every day for about 20 minutes and gets about 15 e-mail messages a day. Two years ago, she gave up her corporate job to have a baby, and now she's doing some real estate consulting on the side.

She's an arts lover and subscribes to two theatre companies, goes to about six arts events a year, and is on three e-mail newsletter lists.

Today, she gets an e-mail from you when she logs on at 10:15 a.m. She's only got 15 minutes before she has to call a client, and she's hoping the baby won't wake up before then.

She and her husband have made plans to go to the theatre on

Saturday night, and she promised to take care of figuring out which show to see and then buying the tickets.

Your e-mail is timely. She likes the discount offer you've sent and clicks over to your site. The last time she looked at your site was five months ago, so she's not really sure she remembers how to navigate it.

She looks around for the ticketing page and quickly scans for words that are meaningful, or navigation that gives her a sense of where to find what she's looking for.

Eureka! She sees a link that says "calendar." Wait, hold on, there's another that says "tickets."

"What should I do?" she wonders, thinking, "Is clicking here going to be worth it? Am I really going to find what I want? Am I going to be confused or frustrated? Maybe I should just call? But I don't have the number....Hmm, maybe a movie is a better idea."

I realize that this scenario might be far from the vision you have of your Web visitors, and it may be a bit exaggerated. But if you design your site with Sabrina in mind, your results will be better for having done so.

Summary Points:

★ How and where you place links, images, and content on your site, and how you organize the information will affect how your visitors use your site.

★ Every time you make your visitors click, a portion of them will choose to leave your site instead.

CHAPTER 8

Design Guidelines

Spend 80% of your time getting the main screen right

The main screen of your site is like the entrance of a building. It sets expectations for the experience inside.

Architects spend a great deal of time trying to envision how people will enter a building, what they will feel at the entrance, where they will go, and what the design of the outside of the building conveys about how the space is organized inside. Your Web site should be designed with the same approach.

When a visitor comes to your Web site, research suggests you've got about seven seconds to tell them a story.

The story needs to start with this thought: What is the site about and what can I do here? You need to present the important information about your organization immediately. That includes not only what the site can do for them, but also something about the feel and tone of your organization.

We begin work with our clients on their site's main screen because it forces them to make a set of decisions about how to organize and present their information, as well as the graphic tone and format of the site.

As you think about your initial design, focus on the following key issues:

- ***Hierarchy of elements***

The size and order of elements on the screen guide the visitors' eyes toward your goal. Rather than making everything the same size, divide up the screen into different sized regions to indicate what's most important.

On the TheatreworksUSA site, the most important feature is helping teachers find shows to which they can bring their students. That goal outweighs all others.

You can see in the illustration below how hierarchy and the size of elements help achieve the Web site's goal.

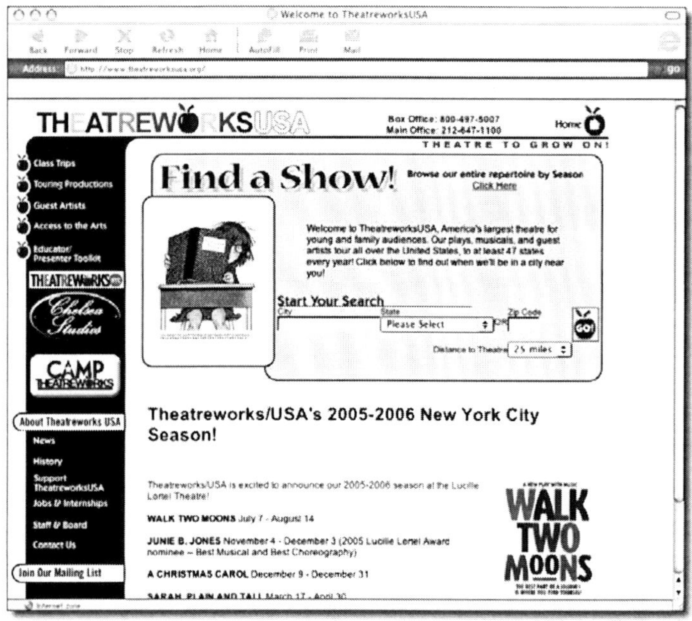

To illustrate this concept further, let's take a look at a before and after view of the Web site for the renowned vocal group Chanticleer.

The organization's old Web site was representative of first-generation arts sites. It was essentially an online season brochure.

In contrast, the new site manipulates hierarchy and subtle imagery to direct the eye to the organization's main goal, which is to sell subscriptions and solicit donations. This appears just under the large image.

Chanticleer's Old Web Site

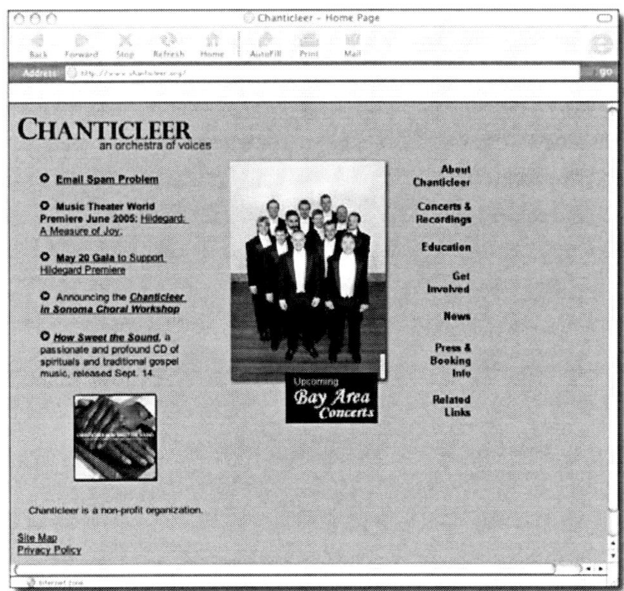

Chanticleer's New Web Site

- ***White space***

It is tempting to fill up every inch of your site with content. However, if you've studied painting, particularly Japanese art, you know that the use of white space on a canvas is often what makes a painting extraordinarily impactful.

Your Web site is no less than a canvas on a computer screen. Your design should use empty space to guide the viewer's eye toward the things that you want them to look at.

The Maine Center for the Arts' site creatively incorporates white space to achieve its design goals:

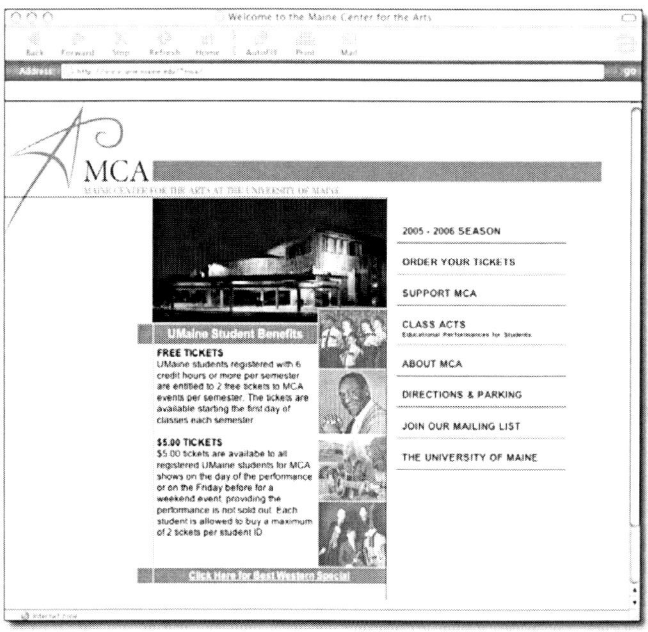

• ***Branding***

The brand of your organization—what your organization's name, image, and tone convey to the public—must be integrated in a meaningful way into your site.

You don't necessarily need to replicate every detail of your branding, but you should maintain the general tone that consumers see elsewhere by using a consistent family of colors, fonts, and imagery. That way, the "look" of any page on the site will immediately identify your organization to the user.

My feeling is that branding is not a sufficient goal in itself, but consistent branding throughout your organization should be an important element in your Web site design.

In this example, the Washington National Opera has branded its 50th anniversary season with a special logo, which appears at the top of the page. This effectively sets the tone for the organization and its special season.

Navigation determines focus

Navigation is to your Web site what the foundation is to a building. It lays out the underlying structure for everything that rests upon it.

In setting up navigation on your site, you will be forced to make decisions about how to group information, which dictates what information will be brought to the viewer's attention most immediately.

Navigation is particularly important since the first thing your visitors will do when they arrive on your site is try to understand the navigation, especially if what they are looking for isn't immediately visible.

Standard navigation on the Web today comes in two basic formats. On some sites, there are links or tabs running across the top of the page (initially made popular by Amazon.com); on others, they run down the left- or right-hand side of the page. If you visit a dozen sites at random, you will probably find these formats used most frequently.

I've found that to create the most effective navigation, there are a few simple rules to pay attention to:

- ***Embrace convention***

Despite the urge to be creative, navigation is an area where I think you should be intensely conservative. Resist the urge to make a creative "statement" with a unique navigational scheme.

The reason for this is that an unconventional approach merely presents a challenge that may frustrate your visitors rather than help them. Since yours is not likely to be a site that your visitors come to daily, you don't have the luxury to teach them a creative new navigational structure that they will become comfortable with through frequent repeat visits.

Generally speaking, I take my cues from what the largest and most highly trafficked sites on the Web do. Below is a list of some of the most frequently visited sites in the U.S., as reported by Nielsen/Netratings.

I urge you to take time to study the navigation schemes on these sites. Become familiar with each of them, focusing on navigation, branding, and use of hierarchy and white space.

Most Visited Sites from Home
(June 2005)

Yahoo.com	83.1 million
Google.com	62.4 million
eBay.com	39.4 million
U.S. Government sites (combined)	31.3 million
InterActive Corp. (includes Ticketmaster & CitySearch)	27.7 million
Amazon.com	26.7 million

In contrast, here's an arts site that breaks the convention. To my eyes, it is unclear where I should focus first, and the navigation offers me too many similar choices, presented in unfamiliar ways.

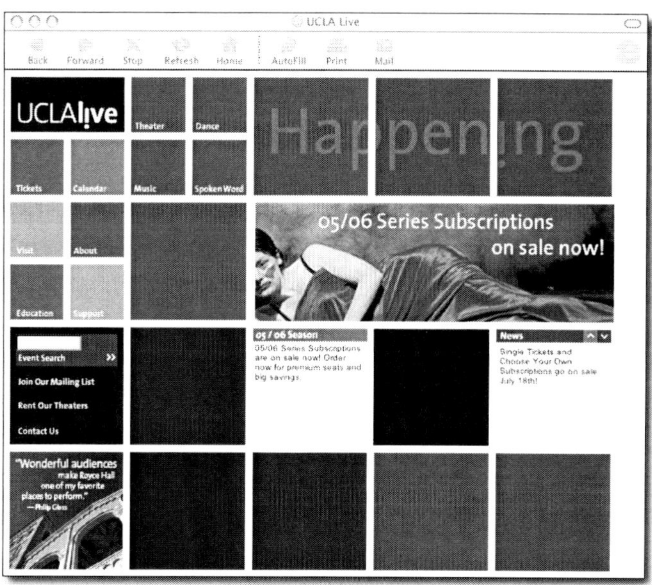

- ***Logical organization reduces frustration***

Making your navigation logical depends on intelligent grouping of information. I've found that the best way to ensure that your site content is laid out in the most logical way is first to create a complete site outline, much as you would outline an essay or article before writing it.

Some sites include this "outline" view of their site's organizational structure on a page called "site map," which can be very helpful to visitors who cannot find what they are seeking from the navigation you've provided.

Site Map: The Laboratory Theatre

Home Page

Visit Us*
- Tickets
- Subscriptions
- Special Offers

Contact*
- Ask a Question
- Address/Phone/Hours*
- Join Our List*

Events*
- Calendar
- Descriptions

Programs*
- Just for Kids
    - o Ages 5 – 11
    - o Ages 12 – 16
- Nationwide tours

Support*
- Become a Member
- Volunteer
    - o Volunteer from home
    - o Help out at the theatre

About Us*
- Staff
- Board
- Company
    - o How we started
    - o Our mission statement
- Location/Hours
- News
- Jobs

* linked directly from main screen

- ***Navigation should be consistent and persistent***

Once you've created a navigational scheme, make sure it appears correctly on each page.

Consistent navigation means that throughout your site each of the navigation links remains the same. I've seen some sites where the navigation suddenly changes between pages on the site. This confuses visitors because they find themselves trying to understand what to them is a new navigation scheme when they just want to find the information they originally sought.

Persistent navigation means that when a visitor moves deep inside your site, the navigation always remains visible. Persistent navigation enables your visitors to move to any other page on the site no matter where they have been. More importantly, it helps them find a familiar page they have already visited on your site, especially useful if they have gotten confused or lost.

If you are integrating pages hosted by a third party (such as a ticketing company), make sure there's a clearly marked link on those pages that will return your visitors to your site, to the exact place where they started.

- ***Offering too many navigational links is overwhelming***

Consumers have trouble deciding amongst vast amounts of data presented to them all at once. For instance, there's now agreement within the consumer package goods industry that offering dozens of brands in one place overwhelms consumers. It is known that consumers buy more frequently when there are fewer choices.

It turns out that too much choice demands too much thinking and ultimately leads to paralysis.

So, when creating your navigational structure, limit your navigation to about 4-6 main sections. Decide on the groupings of information that are most important, and organize all the information on your site into those categories.

- ***Section names should be obvious***

Each of the section titles within your navigation must be in language that is meaningful to a first-time visitor. Use words that are commonly understood as opposed to insider terms only subscribers who are familiar with your organization will know.

Here's an example I'm particularly sensitive to. We work with many clients who have heeded my advice and put a prominent link on their main screen for visitors to sign-up for their e-mail newsletter.

However, when I see a link called "Join our E-Patron Club," I'm confused. The word "club" might imply that I'll be asked to pay a fee or to attend meetings when I sign up.

Thus, some patrons may not click on that link, thinking it's not a newsletter at all. I'd rather see a link that says "Newsletter Sign-Up" or "Click here to sign up for e-news," since I'm pretty certain I know what those phrases mean.

So, review your site navigation to make sure all your terms are completely descriptive and clear.

Reserve space for changing content on the main screen

No matter what central goal you decide upon, there will inevitably be special events, news, and other periodic content that you will want to feature on your main screen.

Reserving a section of your main screen for changing content is wise because it will give your site administrator an easy way to highlight these changes without disrupting the hierarchy of your site or diminishing the quality of the overall design.

Use pop-up pages wisely

Pop-up pages can serve a very useful purpose, despite their unfortunate reputation.

Consumers don't like it when irrelevant pages pop up, because this interrupts them as they browse for other information. As a result, an entire industry of "pop-up blocker" software has developed.

Notwithstanding this, pop-up pages can play a very useful role when they are used as "explanation" pages within a site.

The value to the visitor is that when a page pops up, they remain on the initial page they were viewing. If the link were to take them to an entirely new page, there's a chance they might get lost. This approach eliminates that worry. Amazon.com does this very effectively with frequent links titled "what does this mean?"

Another example of the effective use of a pop-up screen is for the sign-up form for your e-mail newsletter. Think of it this way. Your visitors are not likely to be visiting your site specifically to join your e-mail list. More likely they are seeking information in connection to a potential ticket purchase.

However, during their visit to your site, if they can be sufficiently motivated by your design to sign up for your newsletter, a pop-up page will work well as it won't take them away from the page of information they were initially looking at.

Banish splash pages

Let me be blunt in order to emphasize my point. **Do not include a splash page on your site**. Patrons hate these pages. They offer no intrinsic benefit aside from massaging the ego of your site designers. For the visitor, they merely delay entrance to your site.

Splash pages also force your visitors to pick up their mouse to enter your site, even before they have received any useful information from you.

I'm sure my own reaction to these pages mimics that of a vast number of patrons. When I encounter a splash page, I'm already frustrated.

I hope you see why this isn't a good way to make a first impression!

Be wary of Flash technology

Flash technology has recently become popular among a small segment of the arts community, for exactly the wrong reason—because it's "cool." Sites designed with Flash are animated and present the user with what feels like an unusual and often entertaining experience.

However, I feel sites designed entirely in Flash can be confusing. This is because Flash programmers have at their disposal a larger design palate than do HTML programmers, and as such, they tend to re-invent navigation schemes and generally re-think the entire use of space on the screen.

The end result is often an entirely unfamiliar experience, which frequently is more frustrating than beneficial. My sense is that until some of the big consumer sites on the Web (AOL, Yahoo, and eBay) start building their sites entirely in Flash, it's not a good idea for arts organizations to venture in this direction.

In the meantime, there are plenty of ways to embrace Flash technology within your HTML-designed Web site. This combination is becoming more and more common. For instance, in a relatively recent change, the vast majority of video news clips on CNN.com are now presented in Flash.

To my way of thinking, this approach offers the best of both worlds to the patron, at least today.

Summary Points:

★ Spend 80% of your time getting the main screen right.

★ As you design, focus on these guidelines: hierarchy of elements, white space, and branding.

★ Navigation determines the focus of your site and lays out the underlying structure for everything that rests upon it.

★ Your navigation scheme should a) embrace convention; b) be consistent and persistent; and c) be logical and intuitive.

★ Too many navigational links lead to visitor frustration.

★ Section titles within your navigation must be labeled in language that will be meaningful to a first-time visitor.

★ Use pop-up pages as instruction pages.

★ Banish splash pages and be very wary of building your site entirely in Flash.

CHAPTER 9

Content Considerations

Continuing with the architectural analogy, the content of your Web site is like the furniture in your building. It might move around a bit, but essentially each piece has a specific function and must work with all the others to be as effective as possible and make sense to the user.

Text content

People read differently online than they do in print. The computer screen is less friendly to the eye, and people spend only a few seconds per Web page before deciding whether to read on or abandon it.

Rather than reading every word, they scan for information, looking for keywords or links that will help them find what they are seeking.

Here are a few considerations about working with online text content:

• ***Copy Length:*** When writing promotional copy, keep it relatively short. Limit paragraphs to no more than 50-75 words.

• ***Highlight Keywords:*** Use bold, italics, underlined text, and color to highlight words you want viewers to focus on. Long pages of text don't work nearly as well online as they do in print.

Graphics and images

Your visitor's eye will be drawn instantly to a graphic on your screen before it focuses on the related text. They will interpret what the graphic means before their brain processes the words.

In some cases, they will never read the words at all. Or worse, they will make a judgment about what the page is saying, which may be contrary to what the words are actually telling them.

Since graphics are so powerful, utilize them to help tell the story, to reinforce a message, or to convey a sense of excitement about an event. In other words, don't use graphics merely as window dressing to fill up a page.

Make obvious information obvious

What may be obvious to you, or even boring, may not be to visitors to your site. Let's invoke the watchwords of journalism to guide us here. Make sure that on the main screen of your site your visitors can quickly figure out:

Who: Who are you?

What: What does your organization actually do?

When: What are your hours or your event schedule?

Where: Where are you? How do I get there?

How: How do you present your art?

Why: Why should I care? What's special about what you do?

Let me emphasize something simple, which I see lacking in many sites: address and travel information. Make sure this information is either on the main screen or no more than one click away. This includes:

- Address and phone: include your venue phone and address as well as your administrative offices.
- Travel information: driving directions, map, parking, and public transit.

Encourage interaction and respond quickly

I'm always amazed at arts Web sites where the goal is to sell a subscription or a season ticket, and yet there's no place on the site for a visitor to easily ask a question.

Online consumers often have questions before they buy. Think of the conversations your patrons have at your box office. How many times is their conversation with your box office representative completely transactional and without questions?

Since your Web site isn't a live experience, you need to offer users frequent opportunities to ask questions. You'll see the results of this approach in the number of sales generated by your site.

Along the way, you can solicit information from them in the course of creating an interactive dialogue. There are many ways to initiate contact with visitors to your Web site:

E-mail address or a "mail-to" link: Displaying an e-mail address on your site that your visitor must manually type into her e-mail software is the least convenient way for her to contact you.

However, you can include a "mail-to" link. This type of link, when clicked on by the visitor, will launch her e-mail software and transfer the e-mail address you specify into her addressee ("To") field.

In both of these cases, your visitor must take the time to write out her question or inquiry. Problems frequently arise because typically people won't provide you with enough information to solve their issue, and it can then take several rounds of e-mail with your staff before you can get enough information to handle their request. This is time-consuming for both them and you, and basically defeats the convenience that the Web promises.

Inquiry forms: An inquiry form is a page on your Web site that offers a series of check boxes and blank fields for your patrons to fill out, soliciting exactly the information you need from them. When you request information this way, your respondents can articulate their needs more clearly to you, and you will then be more likely to service that inquiry effectively and quickly. Some forms that we have built for clients route the inquiry to different people within the organization depending on which boxes were checked off.

Filling out the form is easy for the patron since she doesn't have to leave your site and open another piece of software. And, once the form is submitted, an automated response can be sent to the patron assuring her that you have received her inquiry and telling her when she can expect to hear back from you.

Here's a good example of a form-driven inquiry page:

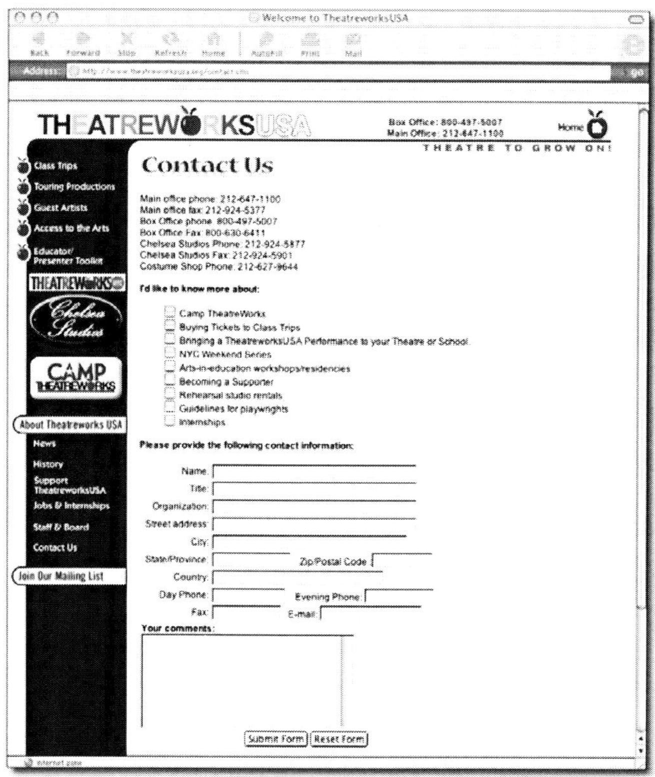

Respond to inquiries quickly

One important way to develop confidence among your Web patrons is to respond to their Web-generated inquiries quickly.

Our rule of thumb is that the visitor should receive a response within one business day with an answer to the specific question asked. Even if a question can't be answered right away, a gatekeeper should respond, indicating that the message has been received and that it has been passed on to the appropriate person in your organization.

When you respond immediately, you convey a sense that your Web site is indeed as valuable a resource as the phone or an in-person visit.

Summary Points:

★ People read differently online than they do in print; they skim and look for keywords.

★ Use graphics to help reinforce a message rather than as window dressing to fill up a page.

★ Make obvious information obvious, such as location, driving directions, opening hours.

★ Begin a dialogue with your patrons by encouraging interaction; respond quickly.

Chapter 10

Site Design Testing

Testing your site is an important element in the design process, and one that is all too often skipped. Indeed, I believe it's the most important (and fun) step of all!

My recommendation is that before you launch your site, but after you have developed the basic design and navigation, you should test it with your patrons to make sure it is working the way you want it to.

I'm going to outline a process for site testing that we've used often. Similar to the research approach I outlined earlier, this is a quick and easy way of collecting information about your site, though not necessarily the approach a Fortune 500 company might use.

Ultimately, this kind of site testing involves nothing more than inviting a handful of people to review your site in person. Here is a basic outline of what to do:

1. Solicit participants

Send a message to your e-mail list to solicit participants. Ask for an hour of their time during a specific time-frame.

Consider offering a small incentive such as a free pair of tickets to encourage participation and to thank them for their help. Sign up at least 10 to 15 participants and assume that about a third won't show up, even after they have confirmed they will come.

If the main goal of your site is to reach new visitors who don't know your organization at all, then you might alter this approach by seeking participants another way, such as partnering with another local arts organization.

2. Administer the test

Arrange for the use of a private office with a computer connected to the internet. Ask your participants to sit at the computer with someone from your staff, but not someone who has been involved in designing the site.

Ask the patron to accomplish a series of tasks on your site, tasks that align with your overall site goals. For instance, if you are a ticket-selling organization, ask the tester to buy a ticket on your site and watch how they go about accomplishing that.

I often find it helpful to ask patrons to narrate their actions out loud. Ask them to describe what they are doing and thinking as they navigate your site. You'll hear an invaluable stream of narration which can reveal important information about what is really going on in their heads as they react to aspects of your site.

Throughout this process, make sure your staff member is neither judgmental nor encouraging. Don't give your patron any sense that they are doing things "properly." Conversely, be very careful not to express any frustration if they do things differently from the way you had hoped they would.

We find that when you collect about 15 responses, you'll begin to see patterns emerge, with the same complaints and praises reoccurring. With this number of responses, you'll have a sufficient sense of the strengths and weaknesses of your design to know what changes are necessary.

If the changes you need to make are significant, you'll want to arrange another round of testing to ensure your fixes have worked.

Finally, you should test your site on various computers with different operating systems and Web browsers, because these factors can cause Web sites to look significantly different. At a minimum you should find computers where you can view your site on both Windows and Apple systems, with the browsers Internet Explorer, Netscape, Mozilla/Firefox and Safari.

Summary Points:

★ Before you launch your site, you need to test it with real patrons to make sure it is working the way you want it to.

★ Testing involves nothing more than watching patrons interact with your site in person.

★ Test your site on different software operating systems and browsers.

CHAPTER 11

Building Your Site

Working with an outside site design company requires as much work as doing it yourself

Our research shows that most arts marketers work with some type of outsider to build their sites. Sometimes this person is a volunteer, a pro-bono agency that a Board member identifies; other times it's a company that has been selected and hired. Some organizations have an in-house staff member who may not work within the marketing department but knows a lot about Web design and can do the programming herself.

No matter which category your organization falls into, this chapter offers tangible steps to guide you in working with an outsider in order to ensure the relationship yields the best results.

Please know from the start that getting effective results by working with an outside design company is not about passing off the work to someone else. I hope you realize that you cannot turn over the key decisions we've discussed earlier to someone outside your organization or your department.

I believe that the role of any outsider is to execute the strategy you've developed. To do this, you must be prepared to collaborate closely with your site designer.

Since my company has been in a position to take over the design work on dozens of sites, I've heard many stories of frustration with outside Web designers or firms. The reason for the trouble is not that these people didn't know their business. Rather, in most cases, neither side had really done their homework before getting started. In brief, expectations on both sides were ill-defined.

Initially, in looking for a Web site designer, you must be clear about the following expectations and goals:

- ***Strategy:*** Are you hiring them to help you develop your strategy or just design your site? Have they agreed to the strategy you've outlined for them?

- ***Design:*** Do you like the design aesthetics and navigation of other sites they have created? Have you reviewed several of their sites carefully?

- ***Hosting:*** Are you asking them to provide site hosting services as well as design?

- ***Industry expertise:*** How well do they really understand your business? I don't believe you should exclude a non-arts industry company from consideration, but you may want to focus on whether there are people at that firm who have interest or experience in the arts and bring a sensitivity and understanding to their work. If you have to spend a lot of time educating the firm about how the arts work, you'll merely be making the project more complex and costly.

- ***Design process:*** Ask them to outline their project plan and process. With regard to design, how many designs will you see initially? What is their process for showing you progress and getting your approval? What happens if you don't like any of their initial designs, and what happens to your price?

- ***Pricing:*** Are you getting a fixed price bid, or will the price vary based on your revisions and/or timetable?

- ***References:*** Talk to more than one of their recent clients. In addition to asking about the quality of the company's technology, ask how well the company met deadlines, responded to requests and engaged in creative problem solving.

Once you've clearly determined what your needs are, I think you'll find the six-step process I've outlined below to be a valuable guide.

1. Assemble your internal team:

If you're going to have more than one person involved in the design of your site, developing an effective team or committee is one of the most important steps in the project.

First, who will you invite to participate? Some organizations have a marketing or Web site sub-committee empowered to get the site built or improved.

It's vitally important that you not only assemble a team (which can include insiders, outsiders, Board members, and even patrons) but that you also create a hierarchy and a review process so that the members understand their role and the extent of their authority.

Second, set the expectations for your team members in terms of their time commitment and, more importantly, their responsiveness. When you're in the thick of working on your site, you are going to need to give feedback to your design agency quickly, typically within 24 hours, if not that same day.

There is no surer recipe for frustration than to give your agency comments on a design they implemented only to go back to them a week later and say, "Well, one member of my group just reported in, and her changes contradict what we told you last week."

At that moment you have frustrated your design agency, and your price has gone up! Make sure your members understand that their participation has to be thoughtful and timely.

2. Decide on authority:

You must decide in advance how you are going to make decisions. Make sure there is a leader, and describe to the group how decisions will be made. Will you have to agree unanimously on a design element? Will there be voting, and if so, how will you vote, and does anyone have a veto?

Ultimately, there must be a single person who has decision-making authority. If you decide everything by consensus, you'll end up with a site designed by committee, which more than likely will be a site filled with confusing compromises.

3. Develop a project brief:

Create a document that outlines your needs. Be as specific as you can.

We help our clients with this process by asking them to fill out a questionnaire, which, when completed, guides us in helping them develop their project brief.

Here is our questionnaire. You'll recognize many themes from earlier in our discussion. This form should be useful to you no matter how you are building your site.

Twenty Questions for Arts Marketers to Answer Before Designing Your New Site

1. How would you rate your current site on a scale of 1 to 5?

2. What is motivating your decision to update your site?

3. What is it that your site lacks now?

4. Is there particular functionality on your site that is missing?

5. Is there anything that is "broken" or terrible on your site now?

6. Can you give the URLs for three sites that you particularly like and say why you like them?

7. What is the overall goal of your Web site?

8. If every visitor who came to your site decided that they were going to click on one link, and one link only, which link would you like it to be and why?

9. Who is the ultimate decision-maker for your new site design?

10. Who else will be involved in design decisions?

11. Who will be responsible for maintaining the site after it is built?

12. Who is providing the content for your site? Does it exist already, or is it being newly created?

13. When would you like to have your site completed? Does that date coincide with a business deadline?

14. About how much have you budgeted for your new site?

15. How much was spent building the existing site?

16. When was it built, and by whom? Who is hosting the site?

17. What was your site's traffic last month?
- Unique visitors:
- Average time on site:
- Average pages viewed:
- Top three pages viewed:

18. How many pages does your current site have?

19. How many pages will the new site have?

20. Which of the following things do you want on your site?
- Online donations
- Online registration
- E-mail list sign up
- Survey
- Chat room
- Message board
- Calendar

21. Are you hiring an outside company to help you with your site strategy or just for design and technology services?

4. Solicit invitations:

When you seek outside help, consider several companies. You'll be amazed at how much you can learn about a firm based on their responsiveness to your inquiry and their professionalism in responding to you.

I recommend that you do your homework and review Web sites of many potential firms. You might interview someone there before actually soliciting a bid.

Then, solicit no more than three actual quotes. Any more than this merely wastes your time and theirs. If you've done your initial homework well, a complete review of three companies will provide you with sufficient choice.

Be straightforward about price. Pricing is always a tricky matter as price does not necessarily correlate to quality. Great Web sites can be built for small budgets. What goes into pricing at individual firms often has to do with their own management and compensation structure rather than the actual work itself.

Many firms will ask what your price range is at the start, and they are right to do so. Just like buying a car, pricing for Web sites is all over the map. So, it is wise to give the firm a sense of your budget, and from that they will glean what kind of client you will be. If you only have $5,000 in your budget, tell them that's your limit.

Some firms may bow out at the start, which is to your benefit. I assure you if you tell a firm your budget range, you will be helping the project more than hurting it. The worst thing you can do is to be silent on this matter and then induce a firm to provide you with a quote for $18,500 when your budget is only $5,000.

It's pretty likely they won't negotiate their price down to your range. And even if they do drop their price drastically to accommodate your budget constraints, you risk an unhappy relationship from the start. This could have an impact on the quality of the site itself. I don't mean to sound alarmist, but don't play games with price.

5. Interviewing and selecting:

I urge you to respond to the firm on the day you receive a proposal from them. A simple thank you indicating that you have received the proposal along with a description of your internal decision-making process and timetable will go a long way toward establishing a good working relationship.

I cannot tell you how many times we've submitted proposals only to be left in the dark for weeks (or months) on end. In some cases we've decided not to pursue a project with a client based on that kind of treatment. You should begin a relationship with each of your prospective site designers as if they are your partners from the very beginning.

All too often, we hear back many months later from an organization that had pushed us for an "urgent" quote. The excuse is invariably, "We were so busy we didn't have time to review your proposal."

This sets up warning bells for us, as it's pretty clear that this client has poor time management skills and will be a problem to work with once the project gets started.

6. Managing the project:

Once you've begun working with your outside vendor, timing becomes very important.

They are going to assemble a team to work on your site, a team that is likely working on other things as well. They will apportion time for your work, time which can only be used effectively if you work closely with them and are responsive to their need for timely feedback.

You will get far better work from a team that is focused on your project and not waiting around for you to respond to design questions.

Summary Points:

★ Working with an outside design company requires as much work as doing it yourself.

★ Developing an effective internal committee is one of the most important aspects of the project, but there must be a single person who has ultimate decision-making authority.

★ Create a document that outlines your site's needs as specifically as possible.

★ Research outside firms and solicit no more than three quotes.

★ Establish a good working relationship with potential firms immediately, and let them know when they will hear from you next.

★ Be ready to work hand-in-hand with your design team when your site is being developed by responding quickly to their need for feedback or approval.

CHAPTER 12

Running Your Site

Once you've completed your site development, running your site on a day-to-day basis involves an entirely different set of requirements than building it. The most important thing to bear in mind once your site is launched is to keep your site updated. This includes but goes well beyond merely keeping your event information current.

Here are some things to pay attention to, all relating to running your site:

- ***Update your site yourself***

Years ago, many organizations developed sites with third-party developers or individuals who "held the keys" to their Web sites. This can make it very difficult, frustrating, and costly to get your site updated.

Technology has evolved such that there is no reason why you should have to rely on an outsider to update your site. There are software products and services that can be installed on any Web site that will give you the power to update your site from your office desktop computer without any knowledge of HTML.

The product we have used and installed on most of our Web sites is called Contribute. It was developed by Macromedia and costs about $100 per user license. I use it myself to update our own Web site and believe me, my HTML skills are not extensive!

A new company in this arena is Edit.com which offers software that provides the same outcome as Contribute, using a Web-based technology approach.

Ultimately, you no longer have to rely on paying a third-party designer or programmer to make changes to your site.

- ***Decide on staffing & authority***

To have a great site, your organization needs to agree on who has final decision-making power for changes and updates to your site.

Assign someone on your staff (maybe that's you) to be in charge of site updates. Their responsibility is to think ahead in terms of what's going on within the organization as a whole so that changes to it are reflected on your Web site immediately.

This should be someone who not only understands how your organization works, but has the authority to make necessary changes immediately.

Finally, develop a timetable and a system by which your site is continually reviewed and monitored. I recommend a thorough review of every page and link within your site at least quarterly.

- ***Update your content***

Online consumers have come to understand that Web sites are more like radio news than newspapers in that they expect sites to be updated frequently. Popular sites with changing content, such as news, entertainment, and financial sites, are refreshed virtually every minute of the day. As such, your arts patrons believe that your Web site is **the most updated place** to find information about your organization.

So, your job is to monitor your site to ensure that it is fresh and accurate. This means that you must remove outdated information instantly and make ad hoc changes on your site no matter how insignificant, to ensure that it is up-to-date. I am sure you understand what a bad impression it makes when you visit a site that prominently advertises an event that happened last night or even last year.

- ***Validate your links***

Just as content must be fresh, so too must your links to other sites. Since the Web is changing constantly, regularly check the sites you have linked to, and make sure the pages still exist and are still relevant to your patrons.

- ***Beware of "feature creep"***

Web sites, unlike almost any other kind of marketing communications vehicle, are flexible, living entities. Because they can be changed on a moment's notice, everyone in your organization thinks they have the right to ask for changes!

Inevitably, once your site is launched, Board members or others in your organization will come to you insisting that their particular issue is so important that you should reinvent your site navigation to put their information in a prominent place.

Ideally, you should have built your site to accommodate changing content on the main screen so that you have an appropriate "slot" to insert changing information.

We often see something many call "feature creep." That's when your site gets littered with additional buttons and navigation elements, or interactive elements that were not part of your initial design goal or navigation but are "slapped on" at the last minute to satisfy a demanding colleague.

If you've done your homework and given authority to one person to manage your site, it will be that person's job to guard against "feature creep."

- ***Monitor Web site statistics***

Once your site has been developed, tested, and launched, I think the real fun of running a marketing-driven Web site begins. Recall what we discussed above about your site being a "machine for marketing." Well, your Web statistics now become the "dashboard" that will tell you how well your machine is working.

If your initial goal was to drive 25% of your page views to a specific section of your site, monitoring your Web statistics will enable you to track your success.

From this point forward, you should track and monitor your Web traffic on a weekly basis. We utilize a simple spreadsheet that tracks which pages on our site get the most visits and what percentage of visitors are accomplishing our stated goal.

Your site hosting company should provide you with software

that summarizes your Web site traffic and statistics. Some of the more popular monitoring programs are WebTrends and Nettracker. This software is not something you should have to buy and install on your computer. Instead, it should be provided to you for free or for a fee by your site host.

Summary Points:

★ To have a great Web site, you need to have an organizational agreement about who has the authority to make changes and updates to your site.

★ Keep your site updated—remove outdated information and events, and make ad hoc changes instantly.

★ Develop a timetable and a system for reviewing and monitoring your site.

★ Update your site using software that allows you to make changes yourself.

★ Guard against "feature creep"—littering your site with new features that don't support your main goal.

★ Monitor your Web site usage to measure your success. Statistics provide a guide for how to improve your site over time and achieve even better results.

CHAPTER 13

Generating Web Traffic

The easiest Web traffic comes by generating visits from those already online

The best place to generate traffic is from those already online. This may seem intuitive, but many marketers spend a good deal of time and money trying to coax someone who is reading a newspaper or listening to the radio to log on to their Web site. They could get far more effective results by first focusing on people who are already online.

Use e-mail to generate visits

The basic premise of effective e-mail marketing is this: if you build a huge opt-in e-mail list and send newsletters and offers to that list, you will drive traffic back to your site.

Therefore, the larger your opt-in e-mail list, the more you will be able to drive traffic to your site. Today, about 3-5% of PatronMail clients' e-mails result in a click from the e-mail to a Web site, in most cases their own site.

Think of it this way: with a 25,000 name e-mail list and a click rate of 4%, a single e-mail newsletter could generate up to 1,000 additional visitors to your Web site in a 48-hour period.

Link like crazy

If people are already online, they are always one mouse click away from your site, if there is a link to click on. It therefore stands to reason that the more places on the Web that have links to your site, the higher your chances are that someone will click on them.

As an integral part of marketing your site, develop a linking strategy and give a staff member or volunteer the responsibility to update and add links on a regular basis. Some of the obvious places you ought to request links from are:

- Local events Web sites (such as citysearch.com and local newspapers)
- Arts festival listing sites
- Chamber of Commerce sites
- Special interest sites—tourism, travel, hotels, children's activities

Getting your site linked is no more difficult than sending a well-worded request to the webmaster or marketing director of these sites. Find an appropriate page on those sites and request to be added there. When sending your letter, it's always helpful to provide the descriptive copy you'd like to have included about your organization, in addition to your URL.

Put a "call to action" in offline marketing materials

I don't believe that merely listing your Web address on your printed materials and in your ads is the best you can do to motivate Web traffic. Just like any action you wish your patrons to take, you must give them a reason to visit your site, a compelling "call to action."

Here are two examples of motivating copy:

- Visit our Web site at www.laboratorytheatre.org for updated events.
- Visit us online at www.laboratorytheatre.org for last minute ticket availability and exclusive discount opportunities.

Search engines

The topic of search engines can get needlessly complex, so I'll keep it simple. Only two search engines really matter, Yahoo! and Google. Together they represent 69% of all searches done online today, according to a report issued by Nielsen/Netratings and Forbes. So, let's focus on how to optimize your site for just these two search sites.

• *Google*

When you search on Google, you're actually accessing a comprehensive index of the Web that Google has created. Google has developed its own software tool, a Web crawler called "Googlebot," which helps Google figure out what's on the Web. Googlebot is really a series of many (many) computers that briefly visit pages all over the Internet on a regular basis, collecting key information and downloading some content. This process is often known as "crawling," as in "Google crawled our site last night."

The words collected from these pages are then indexed alphabetically, and that's really what Google's search engine looks through when you do a Google search.

So as a Web site manager, how do you get yourself included in Google's index? Basically, first by launching your site, and then, more importantly, by letting others know you exist. The Google robot finds new Web pages by crawling the links on the pages it already knows about. Once your site is launched, Google will find you if your site is linked to others.

There is a way to submit your site to Google (http://www.google.com/addurl/), but it's not necessary and it doesn't guarantee you faster placement. Due to the high volume of submissions, you're likely to be included sooner via Google's own automatic crawl than by your submission.

Google's site says: "The best way to ensure that Google finds your site is to have pages on other relevant sites link to yours. Google's robots jump from page to page of the Web via hyperlinks,

so the more sites that link to your pages, the more likely it is that we'll find them quickly."
(http://www.google.com/webmasters/1.html)

Google does a "shallow crawl" of its indexed pages every day to stay updated. They also do "deep crawls" of the entire Web so the index is fully updated about every 6 to 12 months.

If you find all of this fascinating, you can find out when your page was last accessed by the Google robot by doing a search in Google in the following format:

"cache:[your site address]"

If I were doing this for my own site I would type into Google's search box:

cache:www.patrontechnology.com

The result of this search will indicate the time and date of Google's last crawl of your site.

Ensuring that you get a good ranking on Google is an entirely different matter. Google's "PageRank" technology is the name for the process by which it generates and orders its results. What is crucial for you to understand is that your ranking on Google is all about links and keywords.

Google counts the number of pages that link to your site. It also considers the rank of those linking pages in order to determine how important your site is. In other words, if CNN links to your site (and CNN is ranked very highly at Google because it gets a lot of traffic), then that link from CNN will be more valuable to your overall Google ranking than a link from your local bookstore's site.

Google also searches for the number and placement of keywords on your site. It uses those keywords to match against the words that people type into its search engine.

So, if you are running a summer dinner theater in Maine and the words "dinner theater" show up, say, five times on your main screen, chances are that when someone does a Google search for "Dinner theater in Maine" your site will figure highly in those search results.

In other words, the search terms people actually use determine your ranking.

Finally, Google's robot can read text links but not images, so the navigation system on your site should consist of hyperlinked text. (This is yet another reason not to build a site entirely in Flash since Google's robot cannot search within Flash content.)

• ***Yahoo!***

Yahoo has two ways of searching: "Yahoo!Search" and "Yahoo!Directory." Yahoo!Search is very similar to Google. Sites are crawled by its software system and indexed automatically according to links and keywords. You can submit your site to the search engine but it requires a fee which, I believe, is not worth the effort.

Yahoo!Directory is a different kind of tool. It is a listing of Web sites sorted by category and is maintained by live human programmers.

To submit your site to Yahoo!Directory, you start at Yahoo's home page (www.yahoo.com) and navigate to the category you want your site to be listed in.

From there, click on "suggest a site" (in the upper right-hand corner) and input your site information into the form. You'll have the option of free registration (with no guarantee of when your site will be reviewed) or a $299 "7-day Guarantee" submission, which will be re-billed on an annual basis. In certain business-oriented categories you have no choice but to subscribe to this service.

Summary Points:

★ The best way to generate site traffic is from those already online.

★ Mailing to your e-mail list offers the best opportunity to drive traffic back to your site.

★ The more places your site is listed and the more links there are to it, the higher your chances are that someone will click through to your site.

★ To get offline traffic to your site, put a directed "call to action" in print or other traditional marketing materials.

★ There are only two search engines that really matter: Yahoo! and Google. Be sure your site is listed properly on both.

CHAPTER 14

Web Sites and Beyond

One of my close friends often says that building a business on the Internet is like "building a ship as we sail it." The tools and techniques we use today are likely to change in a matter of months as hardware and software improve and the creative adoption of this new technology changes the consumer experience online.

So, as we look ahead to the future, let me focus on a few trends that I think are promising for arts marketers. These are online developments that may become part of your online marketing in the not-too-distant future.

Blogs

Blogs are simply a type of Web site—a public diary typically written by a single person. Blogs can be stand alone sites, or they can be incorporated within a larger Web site. They can be published as an article on their own, or they can be interactive. That is, readers can respond publicly back to the author within some blogs.

I see great value in the blog for arts institutions as a way of enabling better interaction between the creative side of your organization (director, star dancer, conductor, or artistic director) and your audience. Imagine how much more interesting your Web site could be if it included running commentary by your artistic leader about the creation of a work or a season.

You can get started "blogging" (yes, that's a word!) quickly and for free using very simple web-based software. Two popular blog software services are located at www.blogger.com and www.wordpress.com.

Podcasts

Podcasts are nothing more than blogs in an audio format. They are digitally created and provided (usually for free) as a download that can be played on a portable music player such as an iPod, or on your computer itself. Today, podcasts sound like radio broadcasts, only they are not live. They are recorded in advance and can be downloaded for you to listen later.

Just like my discussion of blogs above, podcasts can also enliven your Web site if they are created by key artistic people and positioned as a way to bring your patrons closer to your organization.

You can read more about how to create a podcast at www.feedforall.com and www.podscribe.com, as well as on apple.com's Web site.

Audio & video

Today, the Web is primarily text and pictures, with increasingly more common examples of "streaming media" content. I believe that the next phase of the Web will create an online experience for the consumer that is a more balanced mixture of text, images, audio, and video.

This has great potential for arts organizations since what we are offering (or selling) is often the unfamiliar. Audio and video can convey something about an arts experience that text and pictures simply can't.

Imagine a 30-second video clip with your director talking about how she envisions the latest production you're mounting. Or perhaps get the conductor or concertmaster of the orchestra to talk about a work that the orchestra is about to play.

Many people are put off by confusion over the digital "rights" for video clips—as if the only thing you can show about your art is an actual clip of the production. Don't be thwarted by the legal complications of protected material. Rather, get creative!

A hand-held video camera backstage can offer a fascinating glimpse into your creative process with the potential to be much richer than a traditional video clip of your production.

Portability

As the miniaturization of technology continues, the cell phone, the portable music player (such as an iPod), the digital camera, and the computer will all morph together into a single hand-held device. The notion of the desktop or even laptop computer as the main access point to the Web for consumers will give way to a mobile digital future.

As a result, consumers will be able to have more immediate access to the content on your Web site, as well as e-mail notices from you. In all likelihood, they will be checking their mobile computing device as compulsively as they use their cell phones today!

CHAPTER 15

Getting It Done

Before I close, I want to speak about technology pricing in general, lest you think I have been describing a utopian e-marketing dream that's out of reach for your budget.

I'm going to give some examples of work we've done recently for arts clients to frame this discussion. We've built complete Web sites for arts clients starting at $7,500. Even some large and complicated sites have cost under $20,000.

At Patron Technology, we offer a suite of online e-commerce tools, including online donation forms, online subscription modules, event ticketing and registration forms, and general inquiry forms that can integrate with existing Web sites, even if we didn't build those sites.

These e-marketing add-ons are typically priced between $500 and $2,500 each. Many other firms offer similar technology. Some charge much more and others somewhat less.

My main message is that none of this is going to require a capital expenditure that is likely to be beyond the reach of even the smallest arts organization. Many of our clients approach Board members for specific donations to fund their e-marketing initiatives, and we are seeing more and more foundations and corporate grants supporting online e-marketing initiatives. Since e-marketing shows such a good return on investment, development officers can now make a strong case for e-marketing grants, and foundations seem to be responding.

Some clients have successfully approached corporate sponsors, arranging to put a logo on the organization's site or in e-mail messages in return for a technology grant. The funding tide is definitely turning in your favor.

The bottom line is that with the right technology partner you can keep your costs down and bring your organization to the cutting edge of e-marketing for a relatively small investment.

Now that we're nearing the end of our discussion, I hope you're able to see that improving your Web site—making it the best marketing machine it can be—is indeed within your reach. I hope that the principles and key points I've made will motivate you to get to work with confidence and conviction.

In conclusion, let me offer a personal comment. Over the last decade, I have found working in the area of the arts and the internet to be a thrilling whirlwind of change and opportunity. We're writing the rules of interactive arts marketing now, and learning by experimentation.

There's no doubt that the next decade will continue to offer the same for all of us, and I hope that this book will motivate you further to embrace online marketing and to dedicate yourself to making sure your Web site and your overall e-marketing strategy is as good as it can be.

Your work today will position your organization for e-marketing success. It will help you build your audience for the future, and it may very well be the most fun you'll have as a marketing director.

So, go enjoy the work! And I hope you'll write me at **gene@patrontechnology.com** to let me know of your success.

chapter

ONE

A Review of Power Electronics Applications

The field of electrical engineering may be divided into three areas of specialization: electronics, power, and control. Electronics basically deals with the study of semiconductor devices and circuits at lower power levels. Power involves the generation, transmission, and distribution of electric energy, and rotating machines. The stability and response characteristics of closed-loop systems is the concern of the control area of specialization.

Power electronics deals with the use of electronics for the control of large power. The power electronics era started with high power tubes like thyratrons, ignitrons, and mercury arc rectifiers. With the advent of power semiconductor devices like SCRs and triacs, power electronics has become very important in the control of large power. The major component of the power electronics circuit is the thyristor, which is a fast switching semiconductor device whose function is to modulate the power in ac and dc systems. Modulation of power can vary from 10W to 100MW by turning the switch on and off in a particular sequence.

The thyristor family is a group of four-layer silicon devices consisting of a number of diodes, triodes, and tetrodes. The most important of the controlled-semiconductor switches for power applications is the silicon controlled rectifier (SCR), which is a unidirectional power switch, and the triac, which is a bidirectional power switch. The following sections give an overview of the application of thyristor circuits in power control and conversion applications.

1.1 AC-to-DC Converter

The thyristor phase-controlled rectifier is widely used for obtaining variable dc voltage from a constant ac voltage and frequency source.

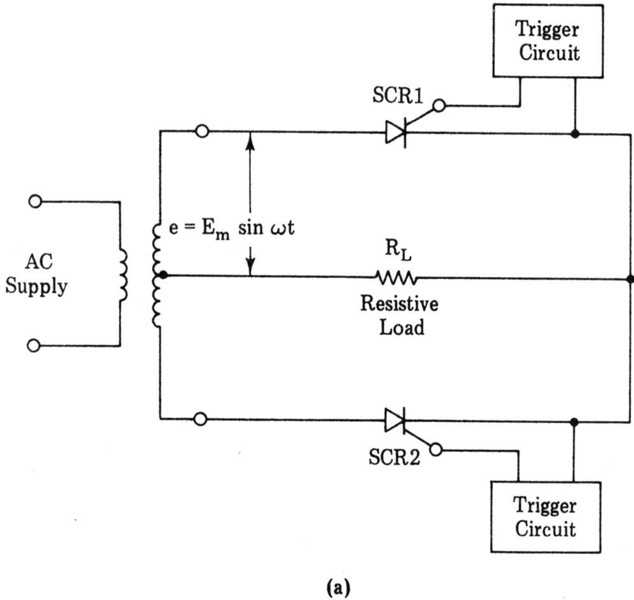

(a)

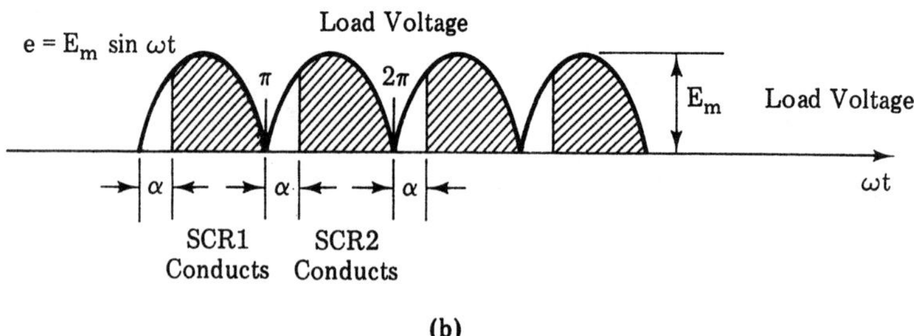

(b)

Figure 1.1. Full-wave centertap phase-controlled thyristor circuit with resistive load: (a) circuit; (b) load voltage waveforms.

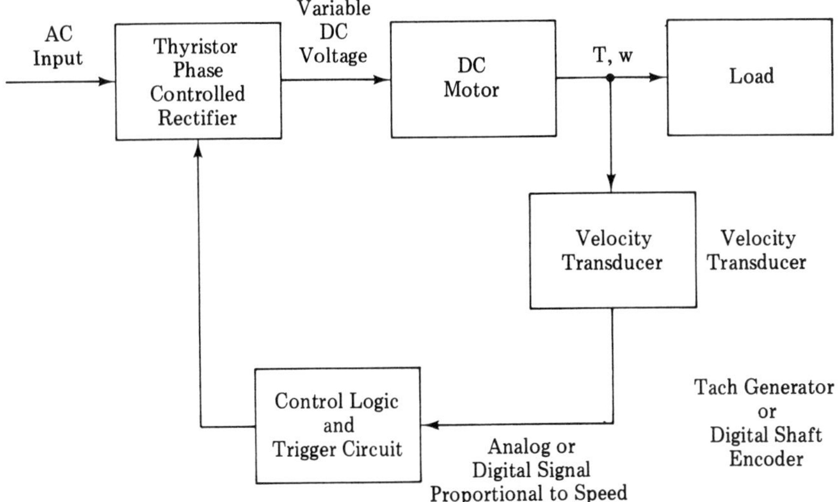

Figure 1.2. Block diagram of a dc motor speed control scheme.

The basic circuit of the phase-controlled rectifier and its principle of operation is given in Figure 1.1.

In Figure 1.1, SCR1 will apply voltage to the load if properly triggered during the half cycle when the top end of the transformer secondary winding is positive (0 to $\omega t = \pi$). Between $\omega t = \pi$ and $\omega t = 2\pi$, the bottom end of the transformer is positive, and SCR2 will apply voltage at the load according to the expression

$$E_d = \frac{1}{\pi}\int_\alpha^\pi E_m \sin \omega t \, d(\omega t) = \frac{E_m}{\pi}[1 + \cos \alpha] \text{ volts}$$

Thus, the output dc voltage is a function of the delay angle α.

$$E_d = 0, \text{ for } \alpha = 180°$$

$$E_d = \frac{2E_m}{\pi}, \text{ for } \alpha = 0°$$

The variable dc voltage from the phase-controlled rectifier is widely used for the speed control of dc motors. Figure 1.2 shows the block diagram of a closed-loop speed control scheme for dc motors using a thyristor phase-controlled rectifier.

1.2 AC Voltage Controllers

AC phase control is used to obtain variable ac voltage from a fixed ac input. Variable-voltage ac output is used for the control of lamps, heat-

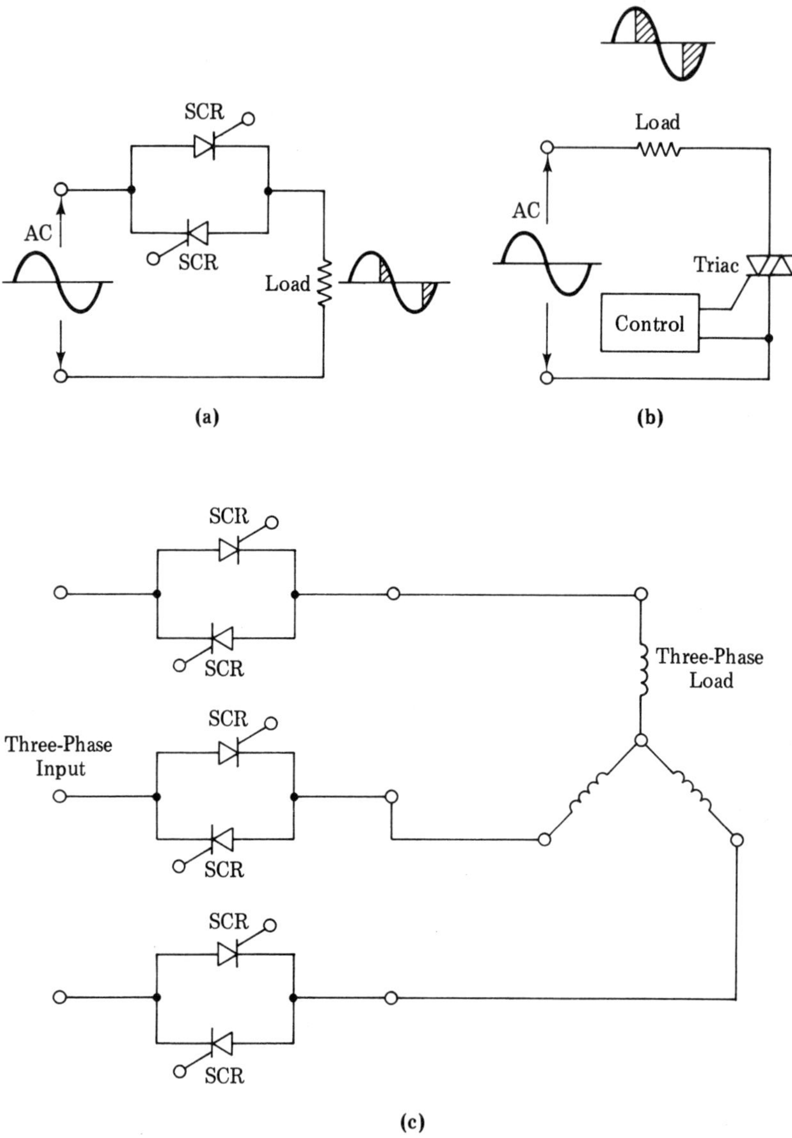

Figure 1.3. Basic ac voltage controller circuits: (a) single-phase ac voltage controller with back-to-back SCR; (b) single-phase ac voltage controller with triac; (c) three-phase ac voltage controller with Y-connected load.

ers, fractional-horsepower ac motors, and three-phase induction motors.

There are two basic methods used in ac voltage controllers:

1. On-off control.
2. Phase control.

AC voltage control circuits are shown in Figure 1.3.

1.3 DC-to-AC Inverters

The dc-to-ac inverters are used to convert the following types of input voltages:

1. DC battery voltage to fixed or variable frequency ac voltage (single-phase or three-phase).
2. AC input voltage rectified to dc, then inverted back to single-phase or three-phase fixed or variable frequency output voltage.

The applications of inverters are the following:

1. Generation of 60 Hz, fixed voltage ac from a dc source, such as dc obtained in wind power, solar generation, or from batteries.
2. Speed control of three-phase induction and synchronous motors.
3. Uninterrupted power systems (UPS).
4. Induction heating.
5. Standby power supply.

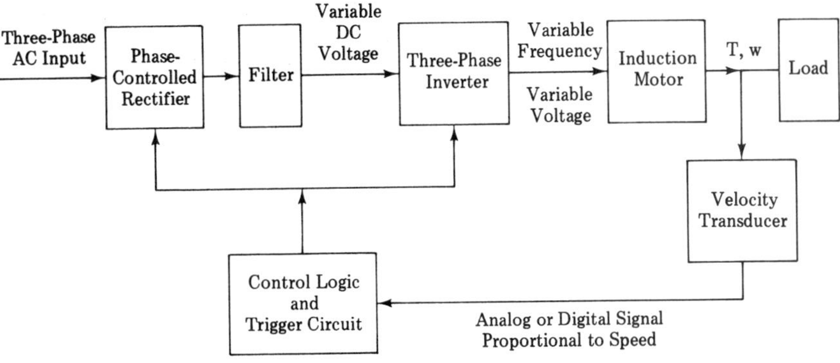

Figure 1.4. Block diagram of a three-phase induction motor speed control scheme.

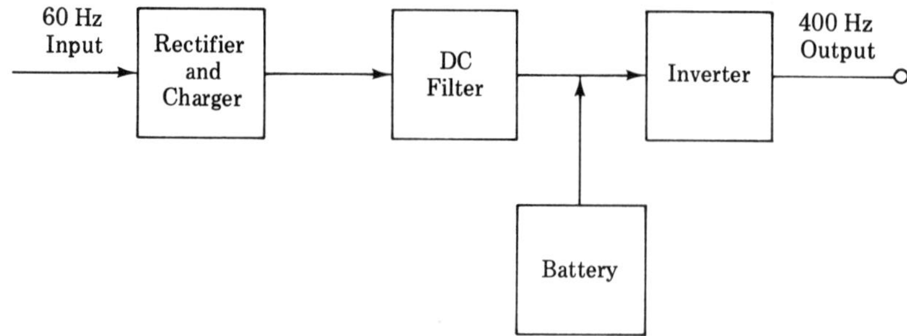

Figure 1.5. Block diagram of an uninterrupted power system.

Figure 1.4 is a block diagram of the closed-loop speed control system of a three-phase induction motor.

Figure 1.5 shows the scheme for an uninterrupted power system.

Figure 1.6 shows the scheme for an uninterrupted power system when the power failure is of longer duration and unlimited protection time is required.

1.4 DC-to-DC Converter

A dc chopper circuit is used to convert fixed dc voltage to a variable dc voltage. A dc chopper is used to control the speed of a dc motor from a battery or dc supply. The block diagram of a dc motor speed-control scheme using the chopper is shown in Figure 1.7.

1.5 AC-to-AC Converter

The naturally commutated cycloconverter principle is used in generating variable-frequency, variable-voltage output from a constant-

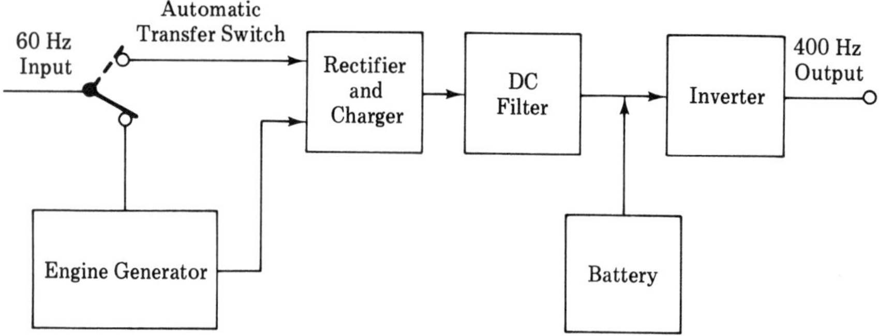

Figure 1.6. Uninterrupted power system with unlimited protection time.

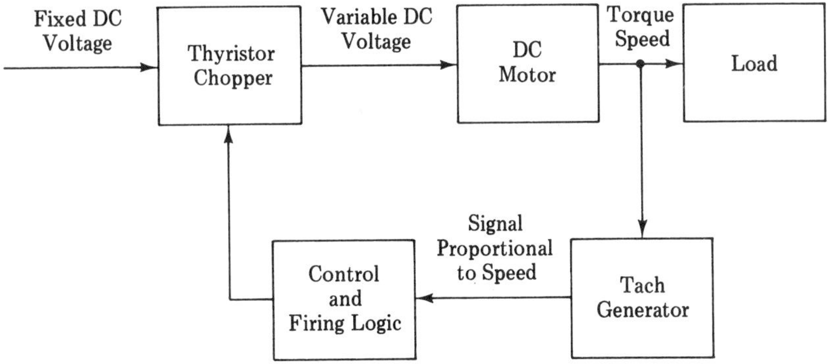

Figure 1.7. Chopper-controlled dc motor.

voltage, constant-frequency ac source. The same principle is used to generate a constant frequency, constant voltage from a variable-frequency ac source.

The cycloconverter principle is widely used in the following applications:

1. Speed control of induction and synchronous motors.
2. Static variable-speed constant-frequency (VSCF) generators for aircraft.

The basic principle of operation of a static ac-to-ac frequency changer is shown in the block diagram of Figure 1.8.

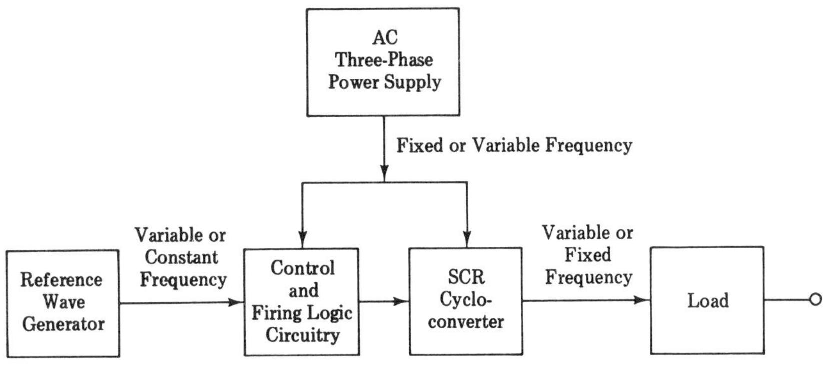

Figure 1.8. AC-to-AC frequency changer system.

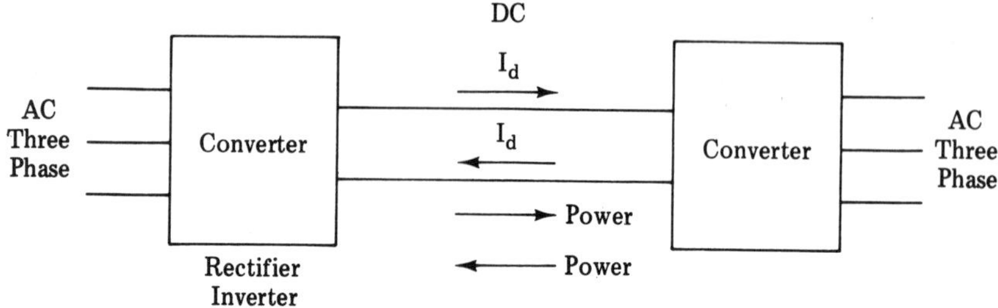

Figure 1.9. HVDC transmission system using line-commutated inverter.

1.6 Line Commutated Inverter

The line commutated inverter is capable of transferring power from ac to dc or vice versa. The direction of power flow in the inverter is dependent on the triggering delay angle of firing of the SCRs in the inverter.

Figure 1.9 shows the block diagram of the HVDC (high voltage direct current) transmission system using a line commutated inverter (converter).

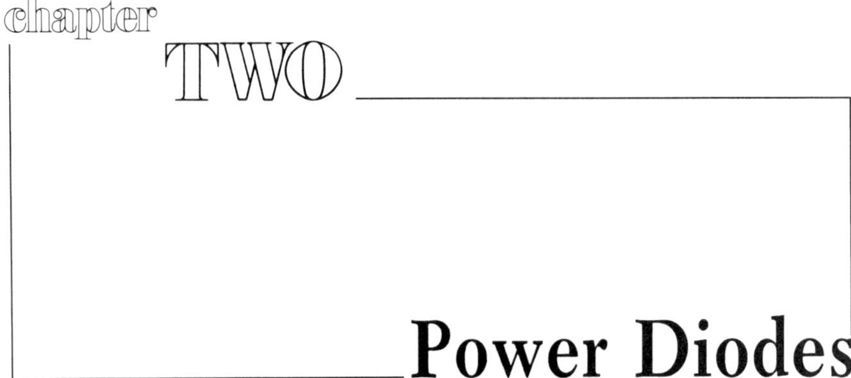

chapter TWO Power Diodes

2.1 Introduction

The power diode consists of a pn junction wafer with one surface soldered to a copper base and the other surface similarly attached to an external electrode. The base forms one electrical connection and the path for heat to flow to an external cooling structure. The base also supports an enclosing insulator through which the other electrical connection passes to the external electrode. The wafer is hermetically sealed from external contamination.

2.2 Construction and Characteristics

The basic symbols and characteristics of power diodes are shown in Figure 2.1. They are similar to pn-junction signal diodes. However, power diodes have larger power, voltage, and current handling capabilities than ordinary signal diodes. In addition, the switching frequencies of power diodes are low compared to signal diodes.

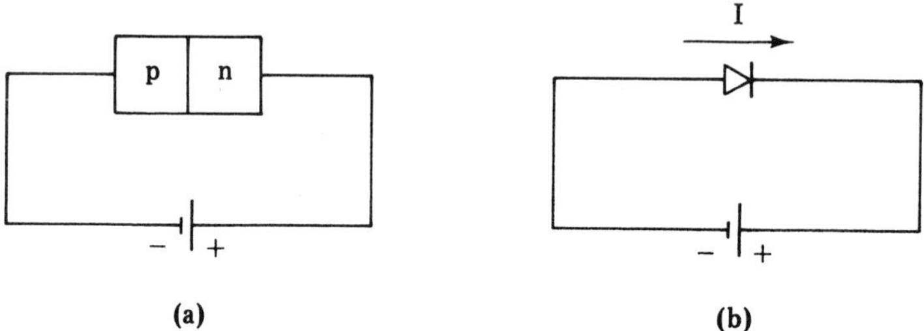

Figure 2.1. (a) A pn junction biased in the reverse direction; (b) The rectifier symbol for the pn diode.

2.3 Ratings

The two most important ratings for a power diode are the peak reverse voltage the diode can maintain short of reverse breakdown, and the maximum steady-state forward current. The diodes are operated near their maximum peak reverse voltages and forward current ratings, as the cost of the device increases as a function of both these ratings.

2.4 Forward Voltage Drop

If the diode is to be used economically, it must be utilized for its maximum forward current rating. Operation under maximum density gives rise to significant voltage drops across the diode. The diode forward voltage drop varies as a function of current and cannot be neglected at high currents.

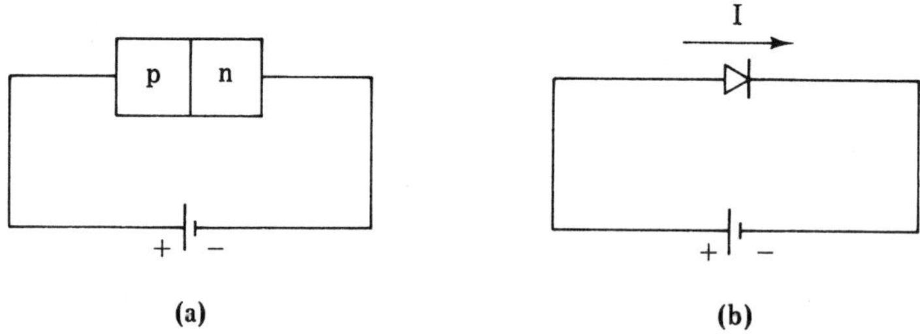

Figure 2.2. (a) A pn junction biased in the forward direction; (b) The rectifier symbol.

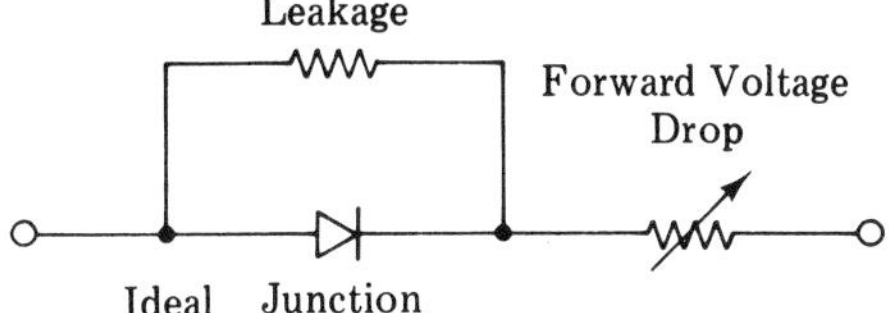

Figure 2.3. The volt-ampere characteristics of a pn diode.

2.5 Leakage Currents

A significant amount of surface leakage current exists around the junction, and is not negligible in high power diodes. Leakage currents become significant in power diodes due to the large fields at the junction when the diode is operated at maximum reverse voltage, and due to the increased area of the leakage path of larger diodes. The steady-state equivalent circuit of a power diode is shown below.

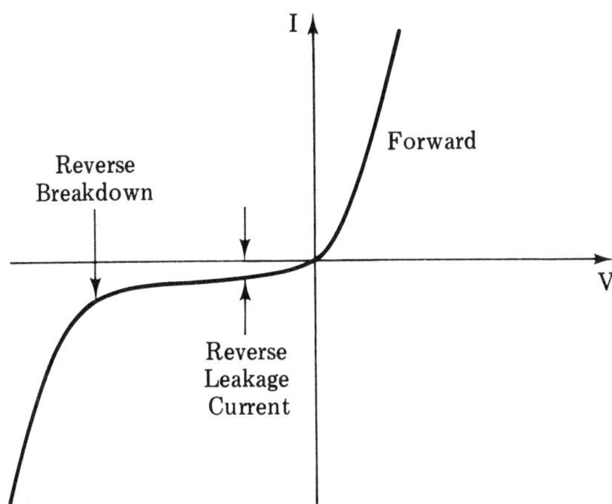

The maximum current surge for a given time is also specified. The only limit on transient forward current is the melting or fusing of the diode or its leads.

2.6 Transient Characteristics

The conductivity modulation effect determines power diode transient behavior. Transient behavior is also influenced by the diode capacitance. The capacitance of the diode increases proportional to the junction area and is therefore proportional to the current rating.

2.7 Series and Parallel Operation

In many applications, the power handling capability of a single power diode is not sufficient. For example, in the converter station terminals of HVDC transmission lines and in rapid transit systems, the power requirements are such that a single diode will not be able to meet them. Series and parallel combinations of diodes are essential for handling the high voltages and high currents necessary in such applications.

Steady-state and transient voltages must be equally shared when diodes are connected in series. When diodes are connected in parallel, each diode must share the total load current equally, both under transient and under steady-state conditions.

2.8 Series Operation of Diodes

In high voltage applications it is often necessary to use semiconductor diodes which are outside the voltage rating range of commercially available diodes. In such cases, diodes may be connected in series to increase the voltage rating.

Consider two diodes (same rating, same type number) connected in series. The V-I characteristics of the two diodes are seldom identical due to production spread. In the forward direction, both diodes conduct the same amount of current, and almost equal voltage appears across each diode. The current rating of the diodes in series is the same as the current rating of one of the diodes.

In the reverse direction, the same current flows in each diode, and each diode supports a different reverse voltage due to dissimilar characteristics. The ratio of the voltages across the diodes will depend on how dissimilar the diode characteristics are, as shown in Figure 2.4. The voltage rating of the diode pair must be larger than the rating of one diode, because part of the total applied voltage will appear across the other diode.

Thus, the series connection of diodes requires either the selection of diodes with identical V-I characteristics or the addition of circuitry to force equal sharing of voltage. A simple and often used solution to this problem is to connect a resistor across each diode. Although a different value of resistor could be placed across each diode to achieve

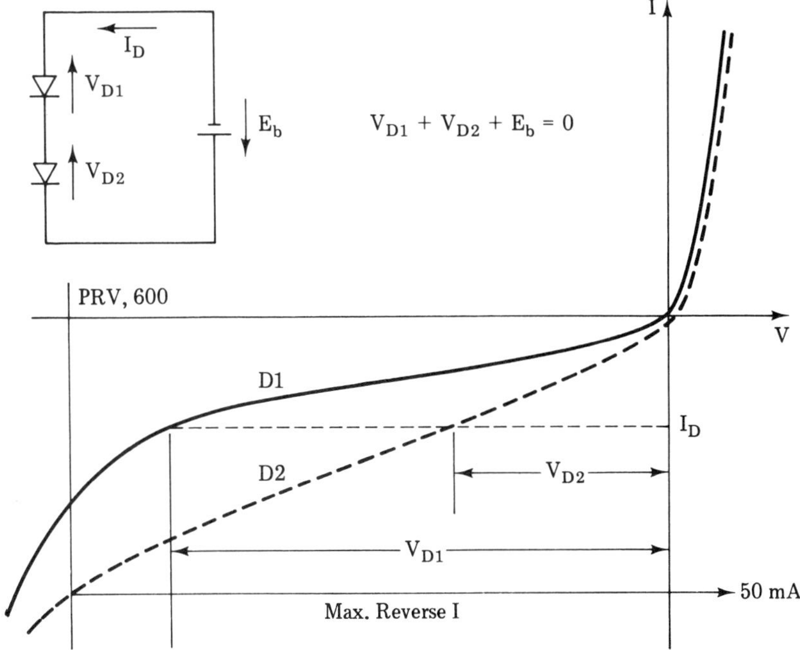

Figure 2.4. Diodes in series.

some optimum voltage division, a more practical approach is to place the same value of resistor across each diode, thereby eliminating the problem of matching resistor values to individual diode characteristics, and making replacement of defective units simple. Figure 2.5 shows the effect of placing resistors across the diodes.

The steady-state voltage distribution along a string of series-connected diodes must be ensured by external circuitry. External circuitry also must provide sharing of transient voltage changes, such as might be caused by switching loads, lightning, or initial application of voltages to the diode circuit. If a voltage transient causes an increase in the reverse voltage applied to a series string of diodes, initially all of the diodes pass a changing current which depends upon the transient amplitude, the circuit load, and the diode characteristics. Due to production spread, some diodes will approach a new steady-state distribution of carriers before other diodes. These faster acting diodes in series attempt to control and block the current associated with the voltage transient. Thus, the faster acting diodes will receive reverse voltage greater than their fair share (the total applied voltage divided by the number of diodes). The voltage across the faster acting diode may well exceed the peak reverse voltage rating of the diode and cause

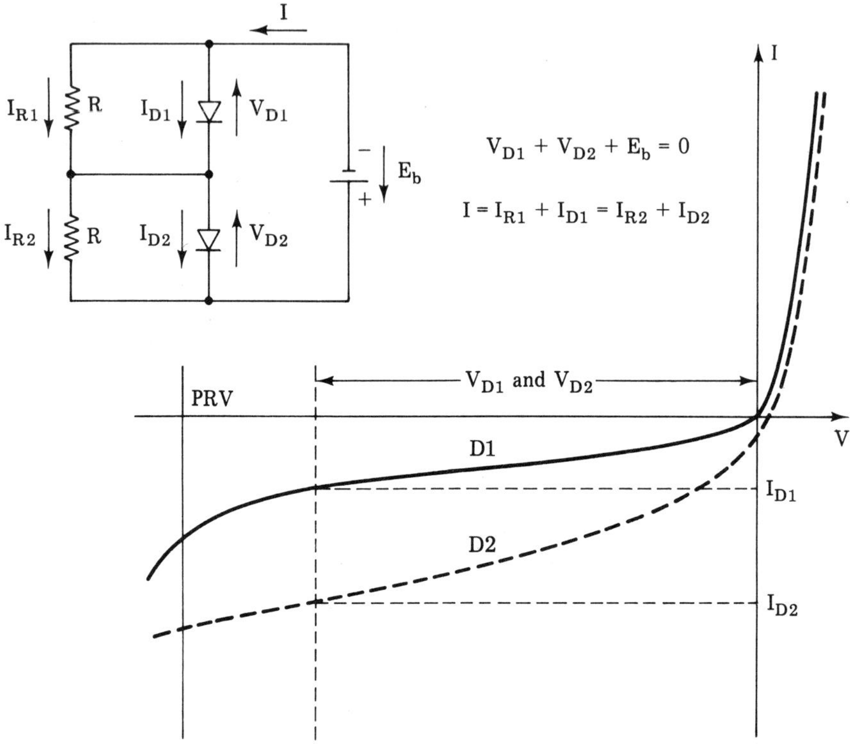

$$V_{D1} + V_{D2} + E_b = 0$$

$$I = I_{R1} + I_{D1} = I_{R2} + I_{D2}$$

Figure 2.5. Diodes in series (resistors added).

the diode to fail. If the diode fails by shorting out, the voltage on the other diodes of the string increases causing other diodes to fail.

Series connected diodes are protected from voltage transients by connecting a capacitor around each diode as shown in Figure 2.6. The capacitors bypass the voltage transients around the diode string, di-

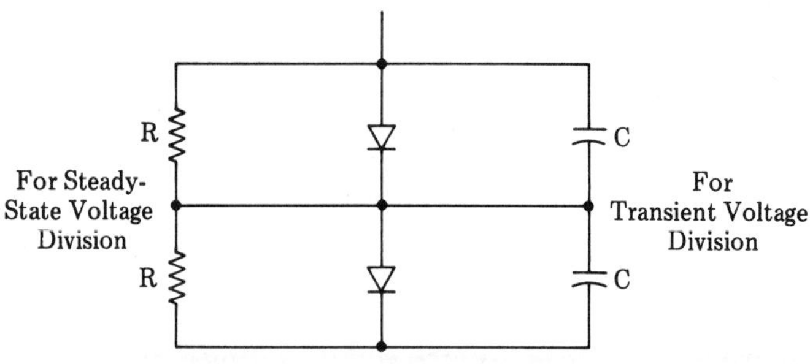

Figure 2.6. Diodes in series.

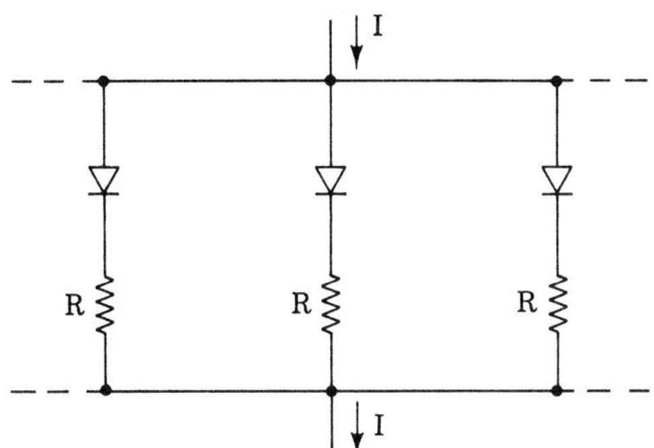

Figure 2.7. Diodes in parallel.

viding the voltage transients equally among the diodes by a capacitive divider action, and limiting the rate of change of voltage across the fast acting diodes.

The expected transient magnitude must be estimated in order to choose the capacitor voltage rating as well as the PRV of the diodes, and the entire circuit must be considered in choosing the value of capacitance necessary to limit the rate of change of voltage across the fastest acting diode.

2.9 Parallel Operation of Diodes

It is necessary to operate diodes in parallel, when load current is very large and a single diode is not capable of handling the load current. Due to production spread, diodes connected in parallel do not share the load current equally. External circuitry is therefore essential. A very small resistance is placed in series with each diode to ensure equal division of current in parallel connected diodes. Transient phenomena are not very important in the case of parallel connected diodes. The peak reverse voltage across each diode is the same. Figure 2.7 shows diodes connected in parallel and resistors in series with each diode to force sharing of current.

chapter THREE Thyristors

The SCR is the most important member of the family of power semiconductor devices. There are other power semiconductor devices of the thyristor family which share the latching (regenerative) characteristics of the SCR. They include the triac, bidirectional diode switch, the silicon controlled switch (SCS), the silicon unilateral and bilateral switches (SUS, SBS), light activated devices like LASCR and LASCS, complementary SCRs, and programmable unijunction transistors (PUT).

3.1 Description of SCR

A thyristor or silicon-controlled rectifier (SCR) is a three-terminal, four-layer pnpn semiconductor device, whose electrical characteristics are similar to those of a thyratron. The three terminals form the anode, the cathode and the gate. The structure of the device is as given in Figure 3.1, and its electrical symbol is shown in Figure 3.2. It was first developed by the Semiconductor Products Department of the General Electric Company in the U.S.A. Its characteristics are made

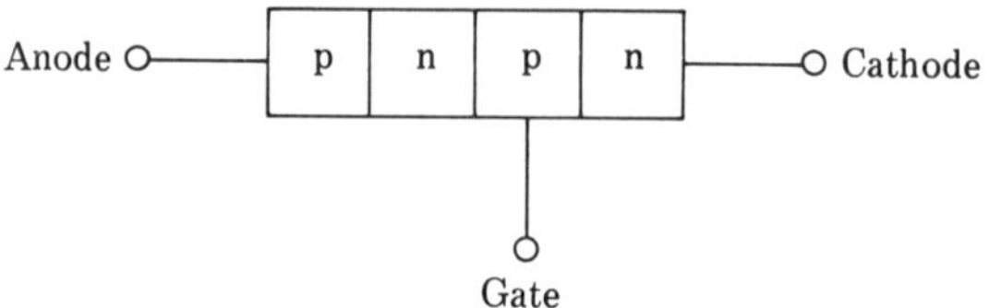

Figure 3.1. Four-layer pnpn structure.

possible through the use of a pnpn junction configuration. Fast switching action, small size, high current, and high voltage ratings make the SCR adaptable to a great many applications.

3.2 SCR Construction

The reliable operation of an SCR depends to a large extent on the design and fabrication of the device. The stud-mounted design facilitates mounting to a heat sink and the top terminal leads facilitate soldered lead connections. The SCR is a disk of four alternate layers of p and n type silicon, a configuration that gives the SCR its unique electrical characteristics. The layers and junctions between them are formed by precision gaseous diffusion and alloying techniques.

The internal construction is shown in Figure 3.3. The most important part of the device is the pnp silicon pellet formed by diffusion techniques. The upper n-region is made by alloying gold-antimony into the p-type silicon. This completes the pnpn structure and gives the device its basic characteristics. The molybdenum disk on the top and bottom are included to minimize the effects of thermal expansion and mechanical stresses on the complete assembly.

The gate or control electrode consists of a small aluminum wire which is connected to the top p-layer of the silicon pellet. This is ohmic contact and not a rectifying junction. Control of the device is accomplished by applying a signal to the top pn junction, i.e., between

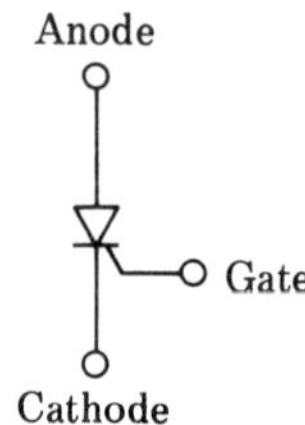

Figure 3.2. Electrical symbol of an SCR.

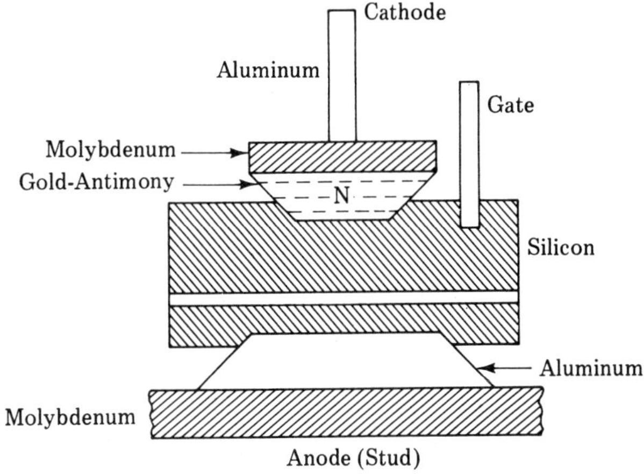

Figure 3.3. Internal construction of an SCR.

the gate and cathode leads. The current alphas of the pnp and npn parts of the SCR must be small in magnitude to have it blocked in the forward direction. In an ordinary three-layer silicon power transistor, it is desirable to have alpha as close to unity as possible in order to achieve a high current gain. Unfortunately, high alphas are obtained in most silicon transistors by using very thin base regions, and a thin base between two low resistivity regions is incompatible with high voltage. The wide base regions in the SCR, necessary to achieve low alphas, are compatible with high voltage so that it is inherently a higher voltage device. An advantage of the SCR over power transistors is the amount of drive necessary for full conduction. In many silicon power transistors, it is necessary to inject up to half an ampere of base current in order to conduct five amperes from collector to emitter. In an SCR the amount of current conducted is dependent only on the external circuit once the device has been triggered. Less than 100 mA is sufficient to trigger the SCR, which can conduct more than 100 amps.

In a transistor, the current at the base triggers it to conduction and the whole base region is affected by it. Therefore, when the base current is removed the transistor returns to the cut-off state. But in an SCR, the gate electrode injects current only to a small fraction of the whole base region, P_2 (Figure 3.4). Hence, once it is triggered to the conducting state, the gate electrode loses control and remains in the conducting state, even when the gate current is removed.

The subassembly shown in Figure 3.3 is housed by welding to the bottom molybdenum disk. When the ends of the terminal tubu-

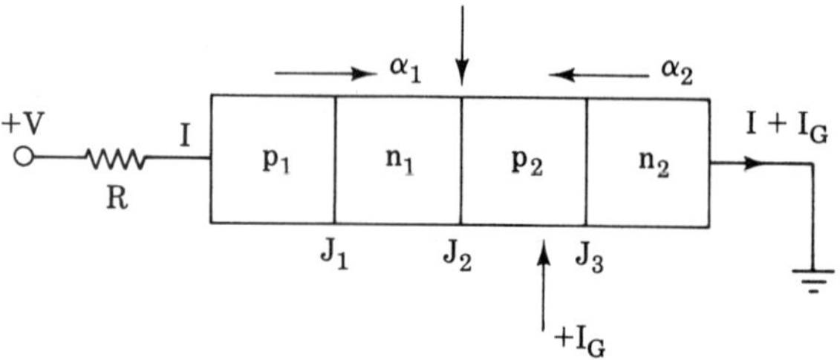

Figure 3.4. Schematic diagram of the SCR.

lations have also been welded, a hermetically sealed unit results. This housed subassembly is then carefully soldered to the copper stud, thus providing a low thermal resistance from the junctions of the cell to the external heat-sink.

3.3 Electrical Characteristics of SCR

The reverse region is typical of any silicon junction power rectifier cell, but the forward characteristics are unique to the SCR.

A voltage V is applied between anode and cathode with anode positive with respect to cathode as shown in Figure 3.5. If the voltage V is below the breakdown potential of the device, only a small leakage current flows. Below the breakdown potential, the cell forward characteristic is actually the reverse characteristic of the center pn junction, and thus provides a very high resistance to current flow. Above the breakdown potential, the device switches from the blocking to the conducting state. Once the device is in the conducting state the forward characteristic is typical of any silicon junction rectifier exhibiting low dynamic resistance to current flow.

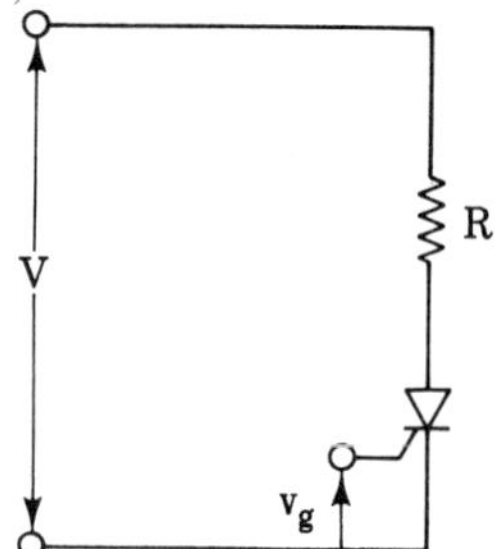

Figure 3.5. SCR operation.

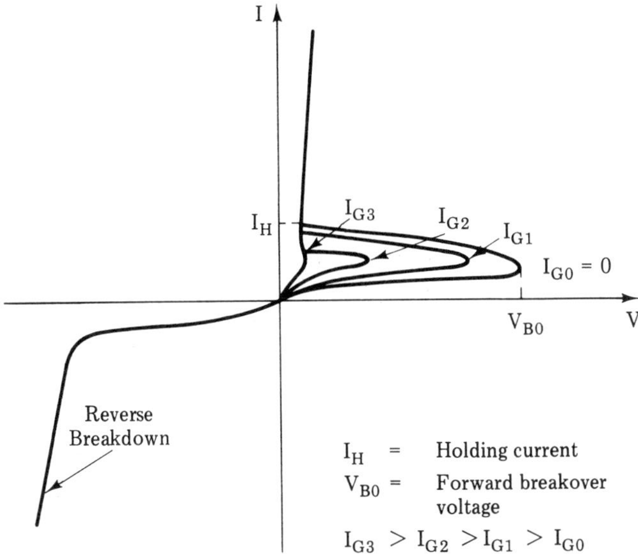

Figure 3.6. V-I characteristics of SCR.

Below the breakdown potential, if the gate is made positive with respect to cathode and gate current flows, the device breaks down and a heavy anode current flows, depending on the voltage V and the resistance R of the anode circuit of Figure 3.5. The V-I characteristics of the SCR are shown in Figure 3.6.

3.4 The Two-Transistor Model of an SCR

A primitive but easily understood analysis, which demonstrates the positive feedback action that results when pnpn device switches, can be carried out with the aid of a two-transistor analogy. The device can be considered as an npn and a pnp transistor connected with the collector of one transistor attached to the base of the other and vice versa, as shown in Figures 3.7a and 3.7b. The base current of the pnp transistor is

$$I_{B1} = (1 - \alpha_1)I_A - I_{cB01} \tag{3.1}$$

However, the collector current of the npn transistor is

$$I_{c2} = \alpha_2 I_k + I_{cB02} \tag{3.2}$$

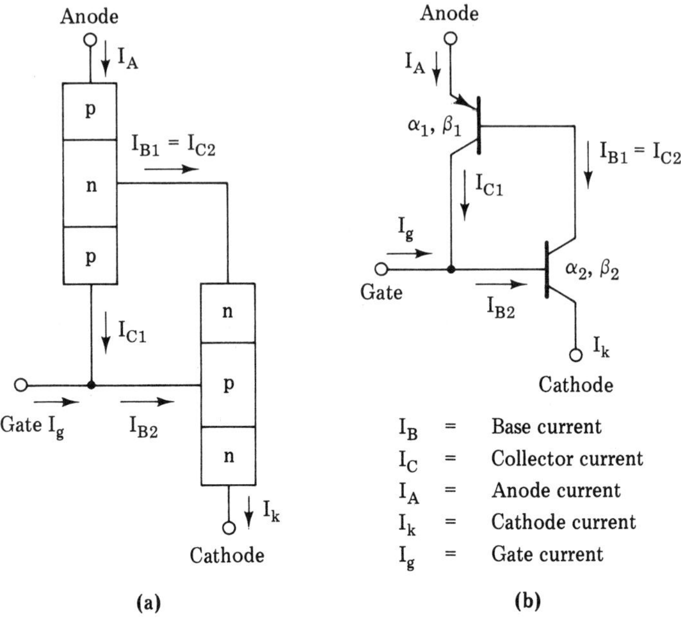

Figure 3.7. Two-transistor analogy of pnpn switch.

By equating I_{B1} and I_{c2}, we obtain

$$(1 - \alpha_1)I_A - I_{cBO1} = \alpha_2 I_k + I_{cBO2} \tag{3.3}$$

But, $I_k = I_A + I_g$, so Eq. 3.3 can be solved for

$$I_A = \frac{\alpha_2 I_g + I_{cBO1} + I_{cBO2}}{1 - (\alpha_1 + \alpha_2)} \tag{3.4}$$

If the base drive current $I_g + I_{c1}$ is equated to the base current of the npn transistor, exactly the same solution results. Thus, we see that a regenerative effect will occur when $(\alpha_1 + \alpha_2) = 1$, i.e., the denominator of Eq. 3.4 approaches zero. With increasing emitter current, the alphas of the device will increase (owing to internal mechanisms), and switching will occur, provided the gate drive raises the emitter-current densities to the point where sum of the alphas is unity. Thus, the SCR turns on when $(\alpha_1 + \alpha_2) \geq 1$. This condition can be satisfied in three different ways as discussed below:

1. If the temperature of the device is very high, the leakage current through it increases, and this increase in current causes the alphas to increase, which may then satisfy the required condition to turn it on.

2. When the current through the device is extremely small, then the alphas will be very low and the condition for breakover can be satisfied only by large values of M_p (hole multiplication factor) and M_n (electron multiplication factor). Near the breakdown voltage V_B of junction J_2, the multiplication factors are very high and the required condition for breakover can be obtained by increasing the voltage across the device to V_{BO}, which will be close to the breakdown voltage V_B of junction J_2.
3. The required condition for breakover can also be realized by increasing the alphas. In Figure 3.4, if a current I_G is injected into the base P_2 in the same direction as the current I across J_2, the current gain of the npn transistor can now be increased independently of V and I, because α_2 depends on $(I + I_G)$ and α_1 still, of course, depends on I. The total current gain will now depend on I_G, and independent means of breaking over is obtained. The presence of gate current modifies the V-I characteristic, as shown in Figure 3.6. This method of triggering is most frequently used in SCR operation.

3.5 SCR Transient Characteristics

Figure 3.8(a) shows an SCR, battery, and resistor connected in series. The SCR is triggered on by a step current into the gate. There is a finite time before the current through the SCR and the voltage across the SCR reach steady-state. While the time required to reach steady-state depends upon the particular SCR in the circuit, times of the order of several microseconds are required. The total time to reach steady-state is subdivided into two shorter times as shown in Figure 3.8d.

The delay time, t_d, is defined as the 10% point of the anode voltage waveform ($.9V_B$) or 10% point of the anode current waveform. The rise time, t_r, is the time required for the anode voltage to drop from 90% of its initial value to 10% as shown in Figure 3.8d. The turn-on time is defined as $t_d + t_r$. Clearly the turn-on time will vary with gate current "drive" and from SCR to SCR. Turn-on times are typically of the order of several microseconds.

3.6 SCR Trigger Circuits

The proper design of firing circuits of an SCR involves three important considerations: 1) choice of a suitable circuit which will supply the firing signal under the required circuit conditions, 2) determination

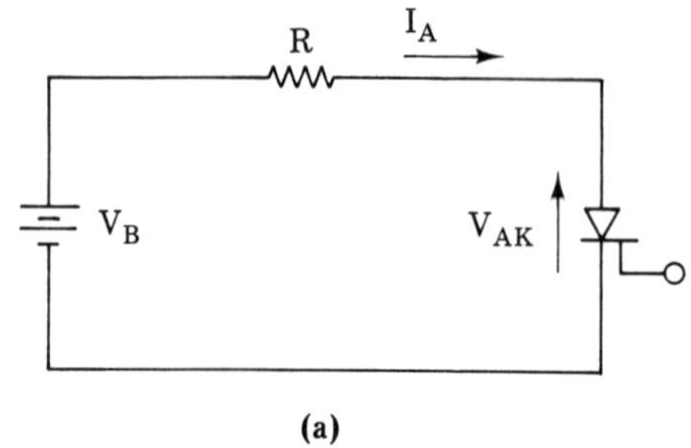

(a)

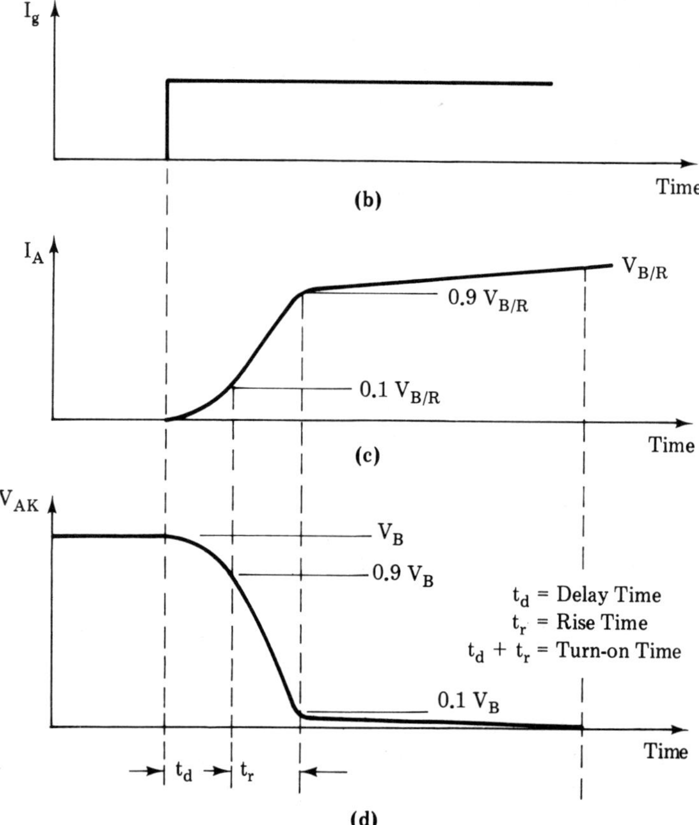

Figure 3.8. Transient characteristics of SCR.

of the maximum voltages and currents which the firing circuit can supply to ensure that the maximum allowable gate ratings are not exceeded, and 3) determination of the minimum voltages and currents which it will supply to the gate to ensure that the SCRs will be reliably fired under all conditions. Less trigger current is required at higher temperatures, while a higher value is required at lower temperatures.

It has also been shown that trigger current requirements decrease slightly as the anode to cathode voltage is increased. Whenever precise timing or phase control is required, it is desirable, due to the temperature and anode effects on the control characteristics and also due to production spreads, to fire with a trigger current of steep wavefront. If the gate signal is a slow variation of dc voltage, or a sine wave, the firing point of the SCR will vary because of changes in the junction temperature. If accurate firing is required, the slowly changing signal must be converted into a relatively fast pulse at a suitable signal level, this pulse being then used to trigger the SCR. Also, the trigger signal must be derived from a low-impedance circuit as it is required to supply appreciable current for firing large SCRs.

Several different circuit configurations may be used to obtain the required trigger pulse for firing SCRs. Before a circuit used to trigger an SCR can be designed in detail, the SCR gate specifications must be considered. Gate specifications may be obtained from a manufacturer's data sheet.

3.7 SCR Turn-Off Circuits

The SCR gate loses control once the anode to cathode current of the device is above holding current. In order to turn the SCR off, the anode current must be reduced below the holding current or the anode made negative with respect to the cathode. If the SCR is supplied from an ac source, the reversal of the voltage during each half cycle serves to turn off the SCR. If the SCR is connected to a dc source or if the SCR must be turned off during a + v_e half cycle of an ac source, some sort of commutation (turn-off process) must be employed.

The current interruption through the SCR for commutation could be achieved by opening the switch placed in two different positions as shown in Figure 3.9.

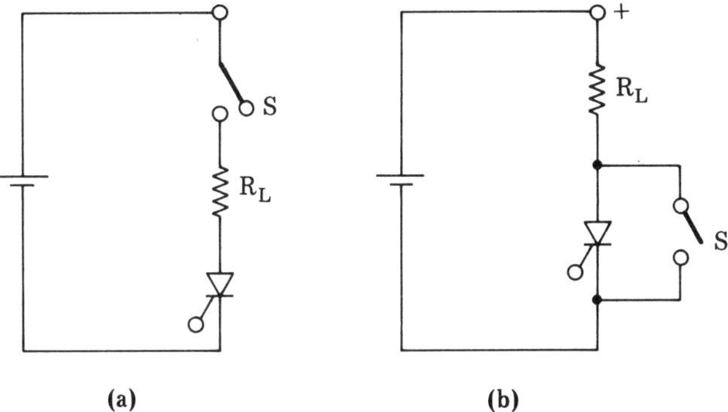

Figure 3.9. SCR turn-off circuits: (a) switch open; (b) switch closed.

The SCR in Figure 3.9(a) is turned off by opening of the switch, S, and in Figure 3.9(b) by closing switch S. After the SCR has been commutated, the switch is opened. These modes of commutation are seldom used except at low power levels.

Applying a reverse voltage to the SCR until it has turned off requires circuitry more complicated than the switch used in the above turn-off circuits. The commutating circuit may be an integral part of the SCR application circuit or it may exist as an auxiliary SCR, functioning only to turn the SCR off. There is a large variety of commutating circuits including (a) underdamped LC circuits which are said to be self-commutating because no other switch device is necessary, (b) capacitor commutated circuits where a charged capacitor is switched across the conducting SCR (the switch is usually another SCR), (c) combinations of the above two schemes, and (d) external pulsed circuits. All of the above methods of commutation are discussed in Chapter 8.

3.8 Triacs and Diacs

A power device with four layers conducts in one direction only. A bidirectional device may be obtained by connecting two of these back-to-back as shown in Figure 3.10. The five layers, n_1, p_1, n_2, p_2, n_3 can be combined into a single structure to form a new device. This is shown in Figure 3.11. When terminal T_1 is positive to T_2 by a voltage

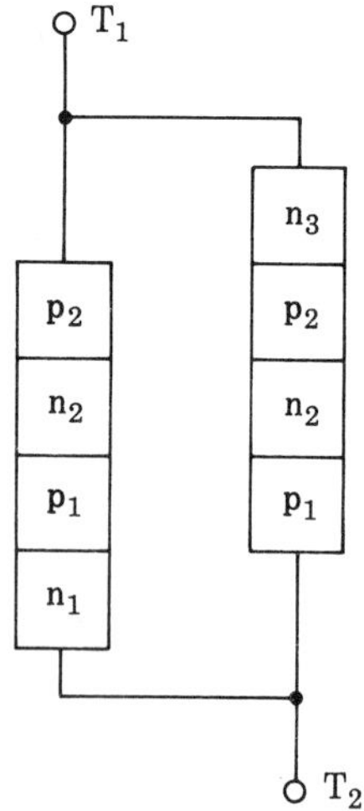

Figure 3.10. Two four-layer devices connected back-to-back.

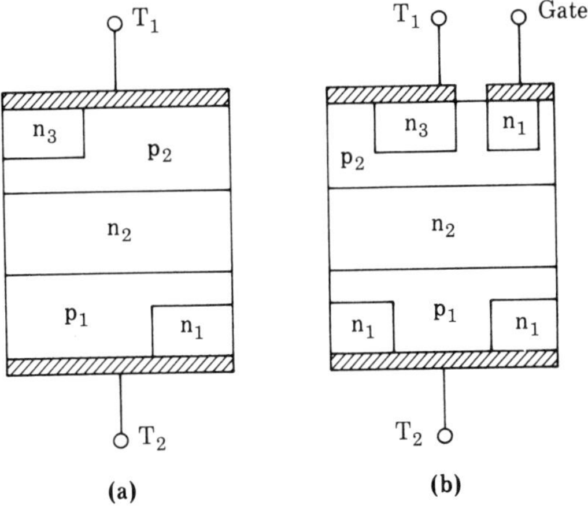

Figure 3.11. Five-layer semiconductor switch: (a) gateless; (b) with trigger gate.

greater than the breakover voltage, the device will break over by normal thyristor action (p_2, n_2, p_1, n_1). For reverse polarity layers p_1, n_2, p_2, n_3 will break over so that irrespective of the voltage polarity, a pnpn structure is presented to the external system.

A five-layer device without a gate can be designed for various breakover voltages and current ratings. A diac is a five-layer gateless device and is shown in Figure 3.11(a). It is used to trigger other semiconductor power switches. Figure 3.12 shows the characteristic for five-layer gateless device; current and voltage ratings are determined by the type of device.

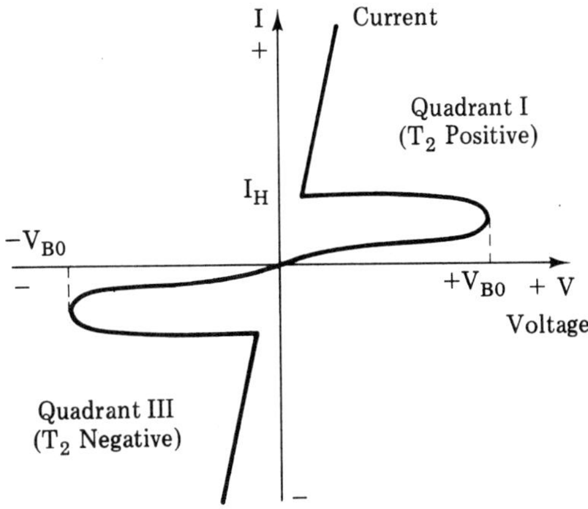

Figure 3.12. Five-layer switch characteristic.

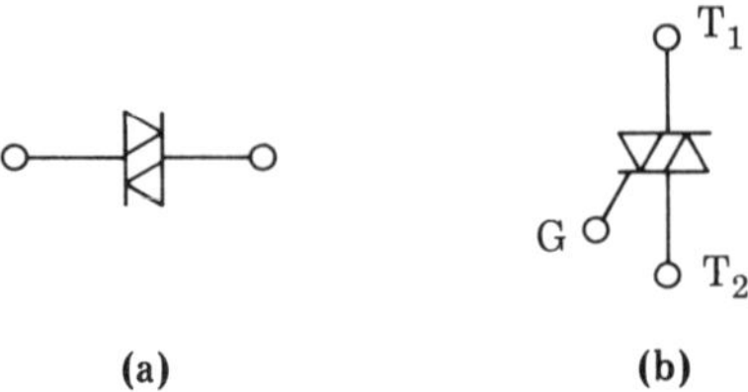

Figure 3.13. Symbols for (a) diac and (b) triac.

A five-layer device without a gate would break over in the first or third quadrant if the voltage exceeds the breakover voltage (V_{BO}). Such a device, called a diac, is manufactured to break over at different voltages and is used to trigger triacs.

A five-layer device with gate is called a triac, as shown in Figure 3.11(b). The V-I characteristics shown in Figure 3.12 can be altered by injecting gate current into or out of the device. The device would then break over at a much lower voltage, similar to that of the SCR. The trigger signal is applied between the gate and T_1.

The triggering modes of the triac are as follows:

1. All voltages are with respect to terminal T_1. When T_2 is positive with respect to T_1 and the gate is positive, the triac is operating in the first quadrant (I^+).
2. When T_2 is positive with respect to T_1 and the gate is negative, the triac is also operating in the first quadrant (I^-).
3. When T_2 is negative with respect to T_1, and the gate is positive, the triac is operating in the third quadrant (III^+).
4. When T_2 is negative with respect to T_1, and the gate is negative, the triac is also operating in the third quadrant (III^-).

The triac is least sensitive in the III^+ mode of operation; this mode should not be used in a practical circuit.

Symbols for diac and triac are shown in Figure 3.13.

3.9 Unijunction Transistor

The unijunction transistor (UJT), complementary unijunction transistor (CUJT), and the programmable unijunction transistor (PUT) are widely used for generation of trigger pulses for thyristors.

The UJT (Figure 3.14) has three terminals which are called the emitter (E), base-one (B_1), and base-two (B_2). Between B_1 and B_2 the unijunction has the characteristics of an ordinary resistance. This re-

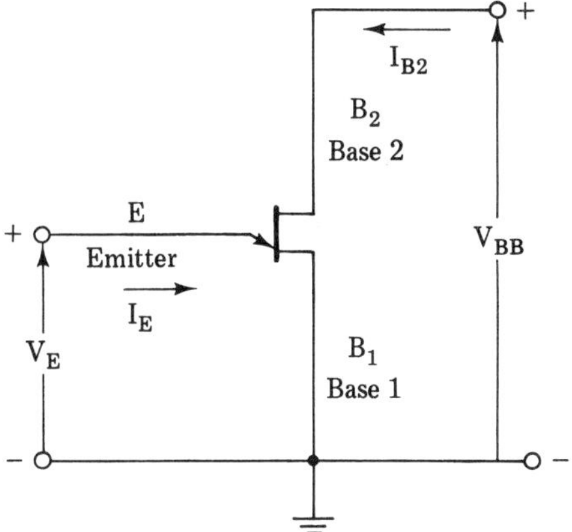

Figure 3.14. Unijunction transistor.

sistance is the interbase resistance (R_{BB}), and at 25°C has values in the range from 4.7 kΩ to 9.1 kΩ.

The normal biasing conditions for a typical UJT are indicated in Figure 3.15. If the emitter voltage, V_E, is less than the emitter peak point voltage, V_p, the emitter will be reverse biased and only a small

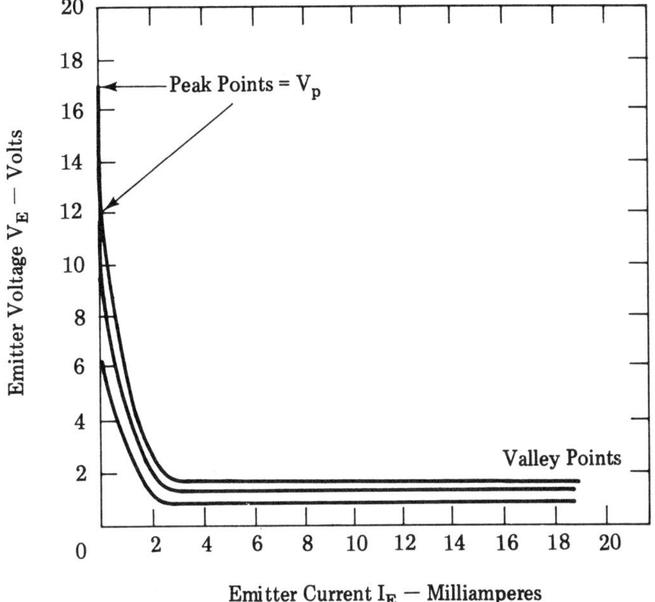

Figure 3.15. UJT biasing conditions.

reverse leakage current, I_{ED}, will flow. When V_E is equal to V_p, and the emitter current, I_E, is greater than the peak point current, I_p, the UJT will turn on. In the on condition, the resistance between the *emitter* and *base-one* is very low and the emitter current will be limited primarily by the series resistance of the emitter to base-one external circuit.

The peak point voltage of the UJT varies in proportion to the interbase voltage, V_{BB}, according to the equation:

$$V_P = \eta V_{BB} + V_D \tag{3.5}$$

The parameter η is called the intrinsic standoff ratio. The value of η lies between 0.51 and 0.82, and the voltage, V_D, the equivalent emitter diode voltage, is in the order of 0.5 volt at 25°C, depending on the particular type of UJT.

3.10 Complementary Unijunction Transistor (CUJT)

Complementary unijunction transistor is a silicon planar, monolithic integrated circuit. It has unijunction characteristics with superior stability, a much tighter intrinsic-standoff ratio distribution, and lower saturation voltage.

CUJT characteristics are like those of a standard UJT except that the currents and voltages applied to it are of opposite polarity. The opposite polarity has been chosen so that standard npn planar passivated transistor processing techniques can be used. CUJTs can be used in most applications that use standard UJTs. Their unique stability and uniform properties make them ideal for stable oscillators, timers, and frequency dividers.

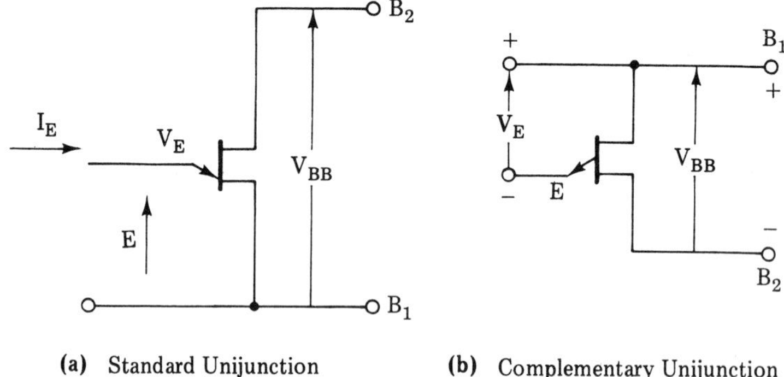

Figure 3.16 (a) Standard unijunction. (b) Complementary unijunction.

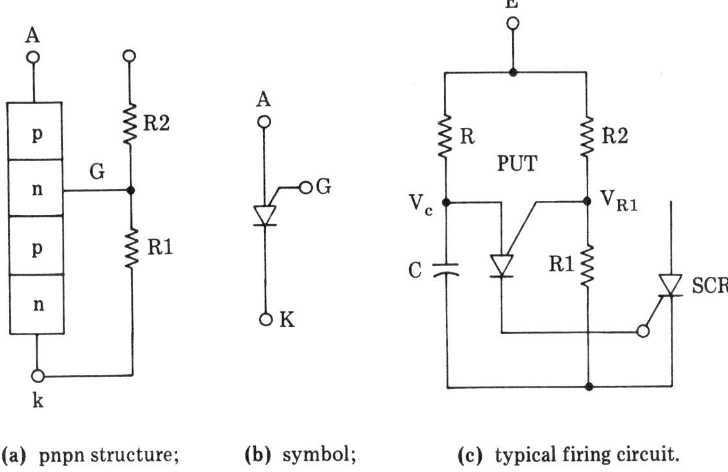

Figure 3.17. Programmable unijunction transistor: (a) pnpn structure; (b) symbol; (c) typical firing circuit.

3.11 Programmable Unijunction Transistor (PUT)

The PUT (Figure 3.17) is a three-terminal planar passivated pnpn device in the standard plastic low-cost TO-98 package. The terminals are designated as anode, anode gate, and cathode.

The programmable unijunction transistor offers many advantages over conventional unijunction transistors. The designer can select R1 and R2 to program unijunction characteristics such as η, R_{BB}, I_p, and I_v to meet his particular needs.

The operation of the circuit is as follows: A definite voltage is applied to the gate of the PUT through the voltage divider action of R2 and R1. The voltage is given by

$$V_{R1} = \frac{E}{R2 + R1} \cdot R1$$

The capacitor is charged to voltage V_c from voltage E through R and C. When the anode voltage V_c exceeds the gate voltage V_{R1}, the PUT starts conducting and triggers the SCR.

3.12 Unijunction Transistor Trigger Circuits

Because of its unique combination of economy, simplicity, compactness, low-power consumption, and high effective power gain, the UJT

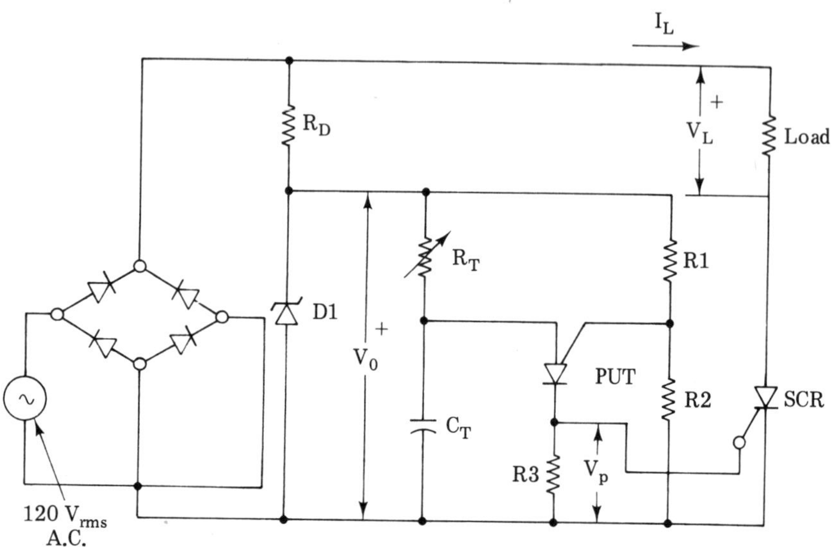

Figure 3.18. Programmable unijunction transistor trigger circuit.

has been widely adopted for a great variety of phase-control triggering applications.

In using the UJT, the circuit designer has a number of basic circuit arrangements from which to choose, each yielding its own specific control transfer characteristics. A single-phase full-wave circuit using PUT as a trigger circuit is shown in Figure 3.18. A similar circuit could be designed using UJT instead of PUT.

3.13 Series and Parallel Operation of SCRs

It was mentioned in Chapter 2 that in high voltage and high current applications, power diodes have to be connected in series and parallel. Similarly, for very high voltage and high current applications, SCRs also have to be connected in series and parallel.

3.14 SCRs in Series

If the input voltage is higher than the voltage rating of the available SCR, two or more SCRs must be connected in series as shown in Figure 3.19.

The V-I characteristics of two identical SCRs are shown in Figure

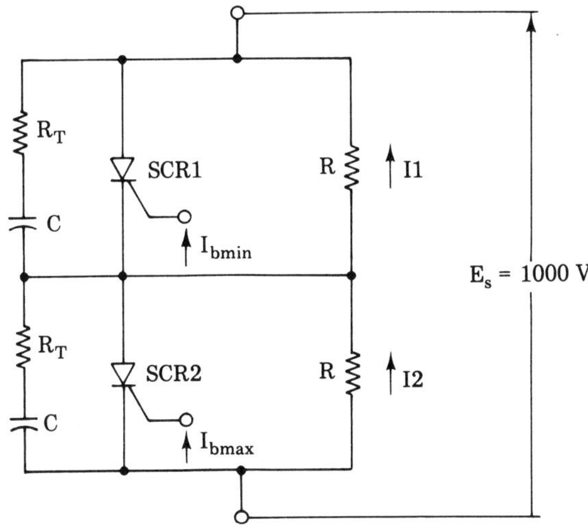

Figure 3.19. Two SCRs in series.

3.20. Due to production spread, the characteristics are not identical. The forward breakover voltages (V_{BO}) are not the same. When two such SCRs are connected in series, the voltage across SCR2 is V2, and across SCR1 is V1. Hence, the two SCRs are not equally sharing the supply voltage E_s = 1000V.

One would expect to block a supply voltage of 2V2. But due to

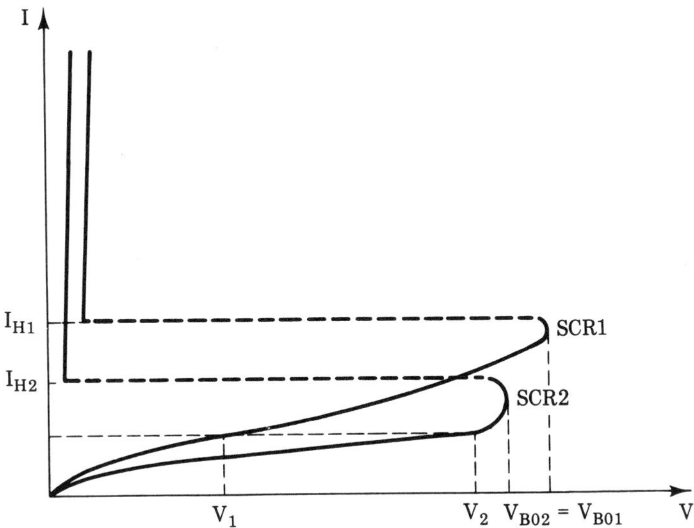

Figure 3.20. SCR V-I characteristics.

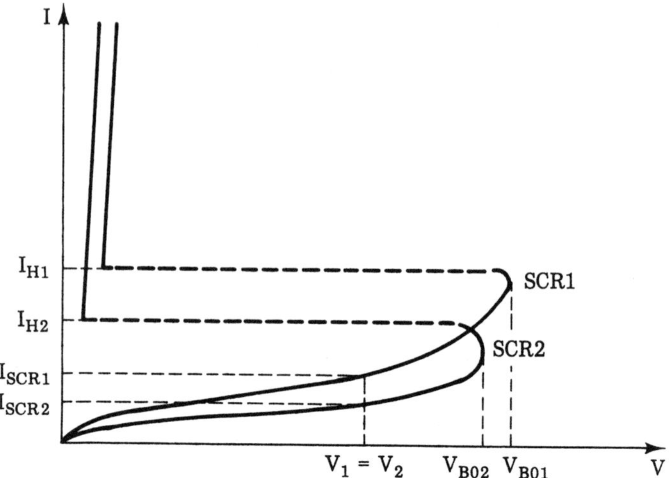

Figure 3.21. Sharing of voltage across SCR1 and SCR2.

differences in the two characteristics, the actual blocking capability is only (V1 + V2). At any further increase in voltage, the V_{BO2} of SCR2 would be exceeded, and this SCR would break over in the forward direction. Hence, in order to force equal sharing of voltages across the two SCRs (under steady-state condition), external resistors (R) have to be connected across each SCR as shown in Figure 3.19. The sharing of the supply voltage is altered as shown in Figure 3.21. Hence in the steady state, sharing of voltage could be accomplished by connecting resistors across each SCR.

The voltage sharing in series connected SCRs could also be achieved by connecting controlled avalanche rectifiers or metal-oxide varistors across each device in a series string.

Example 3.1. *Calculation of Resistor "R" Across Series Connected SCRs for Equal Sharing of Voltage*

The forward and reverse leakage current through the SCR varies due to production spread. The worst condition of voltage sharing takes place in two SCRs connected in series, when the forward or reverse leakage currents of the two devices are furthest apart. The above mentioned condition takes place as seen in Figure 3.19, when SCR1 has I_{bmin} (leakage current), and SCR2 had I_{bmax} (leakage current). The range of leakage current variation is $\Delta I_b = I_{bmax} - I_{bmin}$. Let us choose E_{max} as the maximum blocking voltage across SCR1. From Figure 3.19, $I_1 > I_2$. Hence,

$$E_{max} = I_1 R$$

Further, $$E_s = E_{max} + (n_s - 1)R\, I_2$$

where E_s = source voltage across the series string

n_s = number of SCRs in the series string

$I_2 = I_1 - \Delta I_b$

In Figure 3.19,

$$n_s = 2$$

Therefore $$E_s = E_{max} + I_2 R$$

or $$E_s = E_{max} + (I_1 - \Delta I_b)R$$

$$= E_{max} + I_1 R - \Delta I_b R$$

$$= 2E_{max} - \Delta I_b R$$

Hence $$R = \frac{2E_{max} - E_s}{\Delta I_b} \tag{3.6}$$

In the manufacturer's data sheet only I_{bmax} is specified. For the worst case, if we assume $I_{bmin} = 0$, then Eq. 3.6 reduces to

$$R \leq \frac{2E_{max} - E_s}{I_{bmax}} \tag{3.7}$$

The power dissipated in the equalizing resistor "R" is minimum when the resistor is large. Hence in calculating R, ΔI_b should be used instead of I_{bmax}.

The transient sharing of voltages is accomplished by connecting capacitor across each SCR as shown in Figure 3.19. The resistor R_T in series with the capacitor is to prevent large discharge current through the SCR during turn-on.

3.15 SCRs in Parallel

When the load current requirements are higher than a single SCR can handle, the devices must be connected in parallel. Due to dissimilarity in the characteristics of the SCRs, the total load current is not equally shared by the SCRs connected in parallel. A small resistance in series with each SCR may be used to force sharing of current in parallel connected SCRs.

SCRs with matched forward characteristics may also be used to share load current equally. The current sharing in parallel connected SCRs can also be achieved by connecting parallel reactors as shown in Figure 3.22.

When the current through SCR1 increases over that of SCR2, an

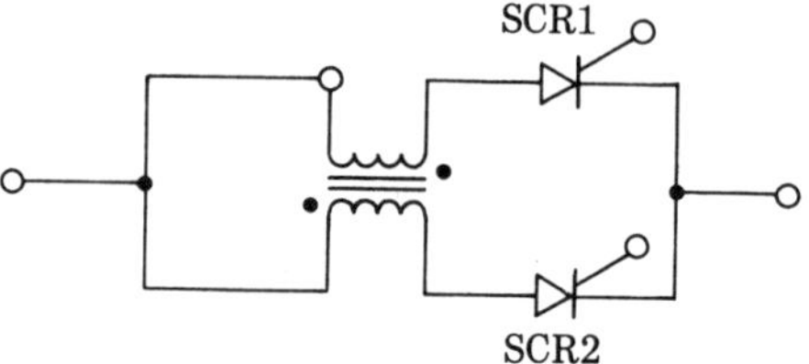

Figure 3.22. Current sharing in SCRs with parallel reactors.

opposing voltage is induced proportional to the difference in current through SCR1. Simultaneously a boosting voltage is induced in series with SCR2 increasing the current flow through the device.

3.16 SCR di/dt Calculation

The maximum permissible rate of change of current through a device is dependent on the type of component being used. The manufacturer's data sheet specifies the maximum di/dt capability. The price of the device increases with di/dt capability. Through proper circuit design techniques, the rate of change of current through the SCR must be kept to a minimum. This section calculates the di/dt through an SCR as a function of other circuit components and the applied voltage. The circuit is shown in Figure 3.23. The circuit voltage equation is given by

$$v = Ri + L\,di/dt \tag{3.8}$$

Solving for current,

$$i = \frac{v}{R}(1 - e^{-t/\tau})A$$

where

$$\tau = L/R$$

$$\therefore \frac{di}{dt} = \frac{v}{R} \cdot \frac{1}{\tau} \cdot e^{-t/\tau} \tag{3.9}$$

or

$$\frac{di}{dt} = \frac{v}{L} e^{-t/\tau} \tag{3.10}$$

The maximum di/dt is at t = 0,

or

$$\left.\frac{di}{dt}\right|_{max} = \frac{v}{L} = \frac{V_m}{L}\ \text{A/sec} \tag{3.11}$$

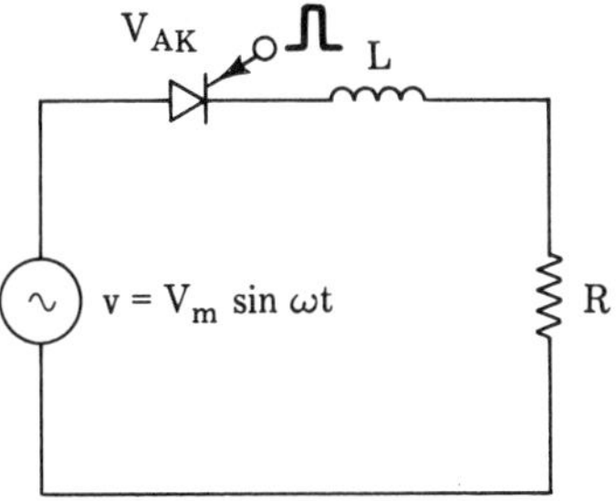

Figure 3.23. SCR di/dt calculation.

It is seen from Eq. 3.11 that the rate of change of current through the SCR is dependent on the supply voltage and inductance in the circuit. The di/dt could be reduced by proper selection of L (inductance) in series with the SCR.

3.17 The Snubber Circuit

The rate of change of voltage (dv/dt) across an SCR has to be kept below the manufacturer's maximum specified value given in the data sheet. The circuit used to limit the dv/dt across an SCR is called the snubber circuit. The snubber circuit is an R-C circuit connected across the SCR as shown in Figure 3.24.

3.18 dv/dt Calculation Across SCR

In practice (as explained in Chapter 8), a dc voltage is switched across an SCR, and the rate of change of voltage across the device must be

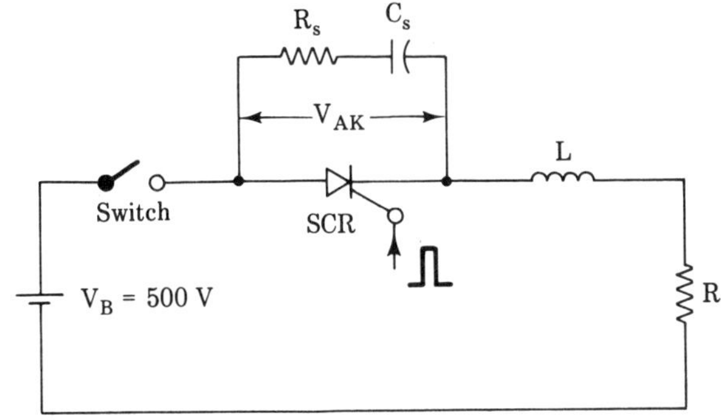

Figure 3.24. The snubber circuit.

below the dv/dt rating of the device, otherwise the device would start conducting even without the application of a gate trigger pulse. The rate of change of voltage across the SCR may be calculated in the following two ways as explained below:

1. When the switch is closed, the voltage equation across the SCR (assuming load resistance R is negligible) is given by

$$V_{AK} = V_B(1 - e^{-t/\tau})V \qquad (3.12)$$

where

$$\tau = L/R_s$$

$$\therefore \frac{dV_{AK}}{dt} = \frac{R_s}{L} V_B e^{-t/\tau} \text{V/sec} \qquad (3.13)$$

The maximum rate of change of voltage is when t = 0,

or

$$\left.\frac{dV_{AK}}{dt}\right|_{max} = \frac{R_s}{L} \cdot V_B \text{ volts/sec}$$

If

$$R_s = \sqrt{\frac{L}{C_s}}$$

then

$$\left.\frac{dV_{AK}}{dt}\right|_{max} = \frac{V_B}{\sqrt{LC_s}} \qquad (3.14)$$

2. When the switch is closed, the voltage oscillation across the SCR is given by

$$V_{AK} = V \sin \omega t \qquad (3.15)$$

where

$$\omega = \frac{1}{\sqrt{LC_s}}$$

$$\therefore \frac{dV_{AK}}{dt} = \omega V \cos \omega t$$

$$= \frac{V}{\sqrt{LC_s}} \cos \omega t$$

The maximum rate of change of voltage occurs at t = 0,

$$\left.\frac{dV_{AK}}{dt}\right|_{max} = \frac{V}{\sqrt{LC_s}} \qquad (3.16)$$

Hence, by proper choice of L, C_s and R_s in the circuit, the dV/dt across the SCR could be limited to an acceptable value. R_s also limits the discharge current through the SCR when the SCR is gated to turn it on.

chapter

FOUR

Phase-Controlled Rectifiers

The thyristor switches allow forward current to be turned on at arbitrary times, but do not permit arbitrary turn-off or commutation without special turn-off circuitry. In normal ac-dc conversion the thyristor could be operated in natural or forced commutation mode. The natural commutated controlled rectifiers are widely used in dc motor control. The natural commutated circuits are easy to design compared to forced commutated circuit. In this chapter controlled rectifier circuits using natural and forced commutation principles are discussed.

4.1 Basic Controlled Rectifier Circuits

The power circuits for the controlled rectifiers are the same whether natural or forced commutations are employed.

In its simplest form, phase control can be described by considering the half-wave thyristor circuit with resistive load R_L shown in Figure 4.1(a). The circuit is energized by a line voltage or transformer secondary voltage, $e_{in} = E_m \sin \omega t$. It is assumed that the peak supply voltage never exceeds the forward and reverse blocking ratings of the

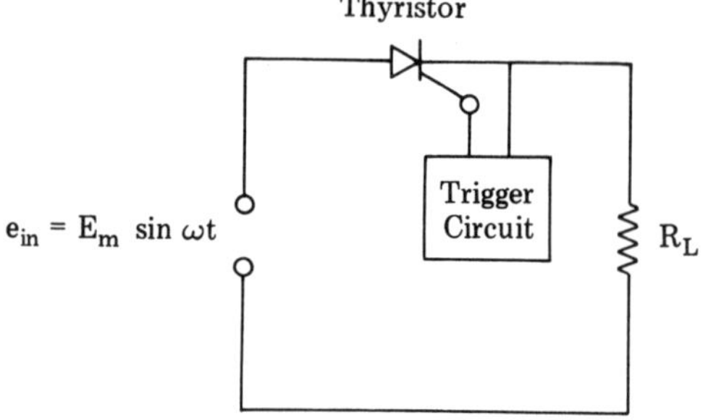

(a) Circuit Diagram and Output Waveshapes with Natural and Forced Commutation with Resistive Load

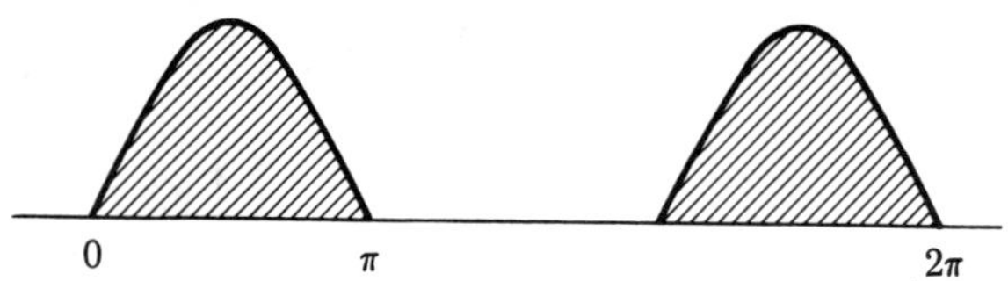

(b) Load Waveshape with No Delay

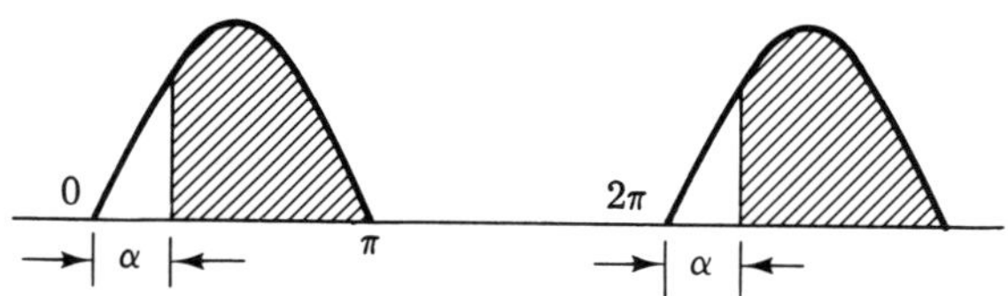

(c) Load Waveshape with Delay Angle 'α' - Natural Commutation

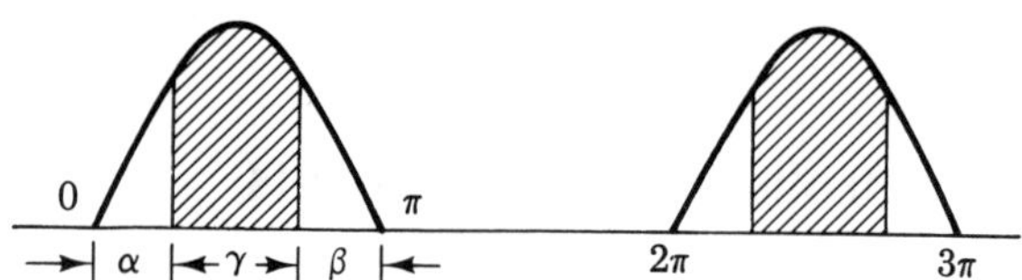

(d) Load Waveshape with Conduction Angle 'γ' - Forced Commutation

Figure 4.1. Circuit diagram and output waveshapes with natural and forced commutation with resistive load: (a) single phase half-wave controlled rectifier circuit with resistive load; (b) load waveshape with no delay; (c) load waveshape with delay angle α—natural commutation; (d) load waveshape with conduction angle γ—forced commutation.

thyristor. During the negative half-cycle of the supply voltage, from π to 2π, the thyristor blocks the flow of load current, and no voltage is applied to the load R_L.

During the positive half-cycle of the supply voltage, the thyristor anode is positive with respect to the cathode, and the gate can exert control over the thyristor conduction characteristics. Until the gate is triggered by a proper positive signal from the trigger circuit, the thyristor blocks the flow of load current in the forward direction. At some arbitrary delay angle α, a positive trigger signal is applied between gate and cathode which initiates thyristor conduction. Immediately the full supply voltage, minus approximately one volt drop across the thyristor, is applied to the load. With a zero reactance source and a purely resistive load, the current waveform after the thyristor is triggered will be identical to the applied voltage wave, and of a magnitude dependent on the amplitude of the voltage and the value of load resistance R_L.

As shown in Figure 4.1(c), load current will flow until it is commutated by reversal of the supply voltage at $\omega t = \pi$. By controlling the trigger delay angle α with respect to the supply voltage, we may vary the phase relationship of the start of current flow to the supply

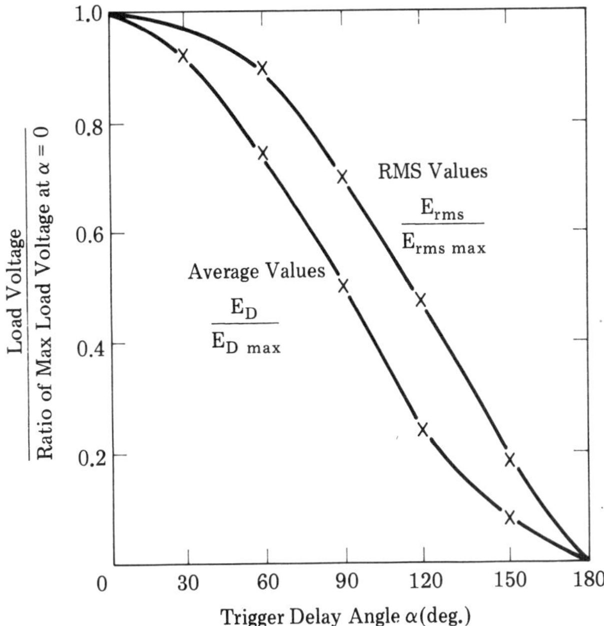

Figure 4.2. Variation of average and rms load voltage with trigger delay angle in half-wave thyristor phase-controlled circuits with resistive load.

voltage and control the load current from maximum value down to zero; hence, the term phase control.

4.2 Forced Commutated Controlled-Rectifier Circuit

In the natural commutation mode, the thyristor is turned off in the negative half-cycle of the ac supply voltage. In the forced commutated circuit, commutation circuitry is necessary to turn off the thyristor in the positive half-cycle of the ac supply voltage as shown in Figure 4.1(d). In this case, it is possible to have a symmetrically-pulse-width-modulated (SPWM) ac/dc converter where $\alpha = \beta$. It is well known that the input power factor of a natural commutated ac/dc converter is poor, especially at low output voltage of the converter. The input power factor could be improved by using forced commutation techniques.

4.3 Average DC Voltage Across a Resistance Load in a Half-Wave Natural Commutated Phase-Controlled Rectifier

$$E_D = \text{average dc voltage}$$

$$= \frac{1}{2\pi}\int_{\alpha}^{\pi} E_m \sin \omega t \, d(\omega t)$$

$$= \frac{1}{2\pi} E_m[-\cos \omega t]_{\alpha}^{\pi}$$

$$= \frac{E_m}{2\pi}[1 + \cos \alpha] \tag{4.1}$$

When the trigger delay angle α is zero, the maximum output voltage is obtained. $E_{Dmax} = E_m/\pi$ (the same output as for a conventional half-wave diode rectifier with resistive load).

With resistive load, the average load current I_D is directly proportional to the average load voltage divided by the load resistance:

$$I_D = \frac{E_m}{2\pi R}(1 + \cos \alpha) \tag{4.2}$$

In some cases, such as with incandescent lamp and electric heating loads, one is interested in the rms value of the load voltage, E_{rms}.

$$E_{rms} = \left[\frac{1}{2\pi}\int_{\alpha}^{\pi} (E_m \sin \omega t)^2 \, d(\omega t)\right]^{1/2}$$

$$= \frac{E_m}{2\sqrt{\pi}}\left(\pi - \alpha + \frac{\sin 2\alpha}{2}\right)^{1/2} \qquad (4.3)$$

The variation of average and rms load voltage with delay angle as expressed in Equations 4.1 and 4.3 is shown in Figure 4.2.

4.4 Half-Wave Controlled Rectifier With Inductive Load

The operation of the circuit on inductive loads changes slightly. Now when the thyristor is fired, at t_{01} say, the load current will increase in a finite time through the inductive load. At t_1 the supply voltage reverses, but the thyristor is kept conducting while the load energy stored during time t_{01} to t_1 is fed back to the supply. The load voltage goes negative, following the reverse half-cycle of the supply voltage. At t_{11} the load current falls to below the holding current of the thyristor and it goes off.

The half-wave circuit is not normally used since it produces a large output voltage ripple and is incapable of providing continuous load current.

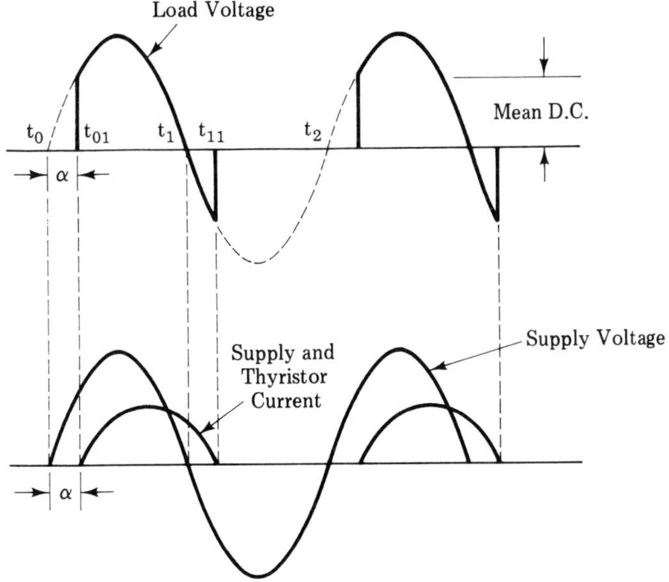

Figure 4.3. Half-wave controlled rectifier with inductive load.

4.5 Single-Phase Full-Wave Circuit

a) Resistive Load

The single-phase full-wave phase-controlled rectifier is widely used in the control of dc motors. The single-phase circuit with centertap transformer is shown in Figure 4.4. The trigger circuit generates firing pulses for thyristors (Th1 and Th2). The trigger pulses are synchronized with the ac supply voltage, and the instant of triggering of Th1 and Th2 can be controlled by varying the delay angle, α, over the full range of 0 to π. Thyristor Th1 conducts when the upper half of the transformer secondary is positive, and Th2 conducts when the lower half is positive. Each half of the input wave is applied across the load. Hence the ripple frequency across the load is twice that of the input supply frequency.

The output dc voltage, E_D, across the resistive load is given by

$$E_D = \frac{1}{\pi}\int_{\alpha}^{\pi} E_m \sin \omega t \; d(\omega t) = \frac{E_m}{\pi}(1 + \cos \alpha) \tag{4.4}$$

Also,

$$E_{rms} = \sqrt{\frac{1}{\pi}\int_{\alpha}^{\pi} E_m{}^2 \sin^2 \omega t \; d(\omega t)}$$

$$= \frac{E_m}{\sqrt{2\pi}}\left(\pi - \alpha + \frac{\sin 2\alpha}{2}\right)^{1/2} \tag{4.5}$$

b) Inductive Load

A single-phase full-wave circuit with inductive load is shown in Figure 4.5. The analysis given here assumes that the inductance is sufficiently large, so that each thyristor conducts for a period of 180° (conduction of current is continuous). Both thyristors are triggered with the same delay angle, hence they share the load current equally. The load voltage and current waveforms are shown in Figure 4.5(b). Due to large inductance in the circuit and continuous current conduction, the thyristors continue to conduct even when their anode voltages are negative with respect to the cathode. The load current is shown to be constant dc.

The output dc voltage, E_D, may be obtained as a function of delay angle, α, as shown below:

$$E_D = \frac{1}{\pi}\int_{\alpha}^{\pi+\alpha} E_m \sin \omega t \; d(\omega t) = \frac{2E_m}{\pi}\cos \alpha \tag{4.6}$$

The output voltage is maximum when $\alpha = 0°$, zero when $\alpha = 90°$, and negative maximum when $\alpha = 180°$.

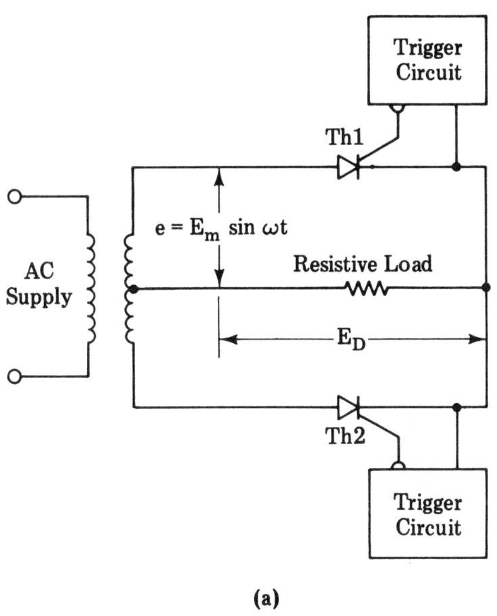

(a)

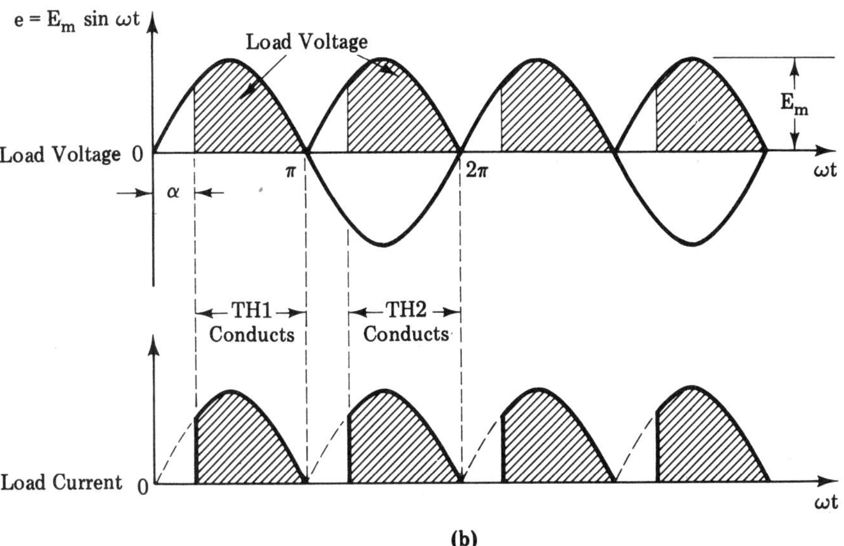

(b)

Figure 4.4. Full-wave centertap phase-controlled thyristor circuit with resistive load: (a) circuit; (b) waveforms.

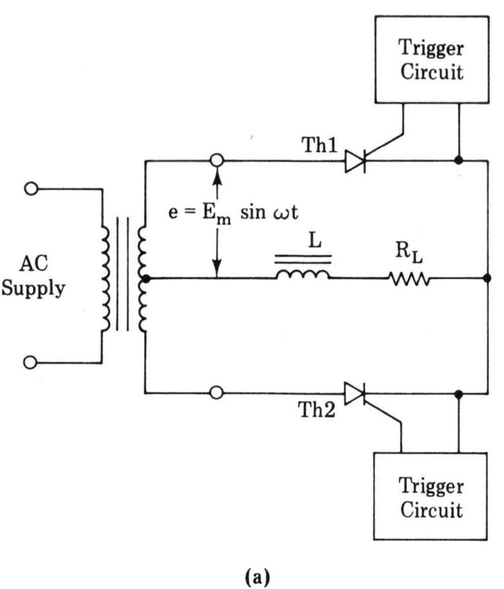

(a)

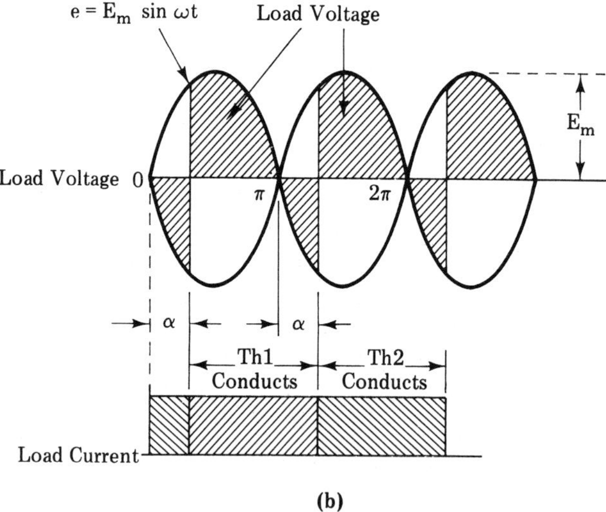

(b)

Figure 4.5. Full-wave centertap phase-controlled thyristor circuit with inductive load: (a) circuit; (b) waveforms.

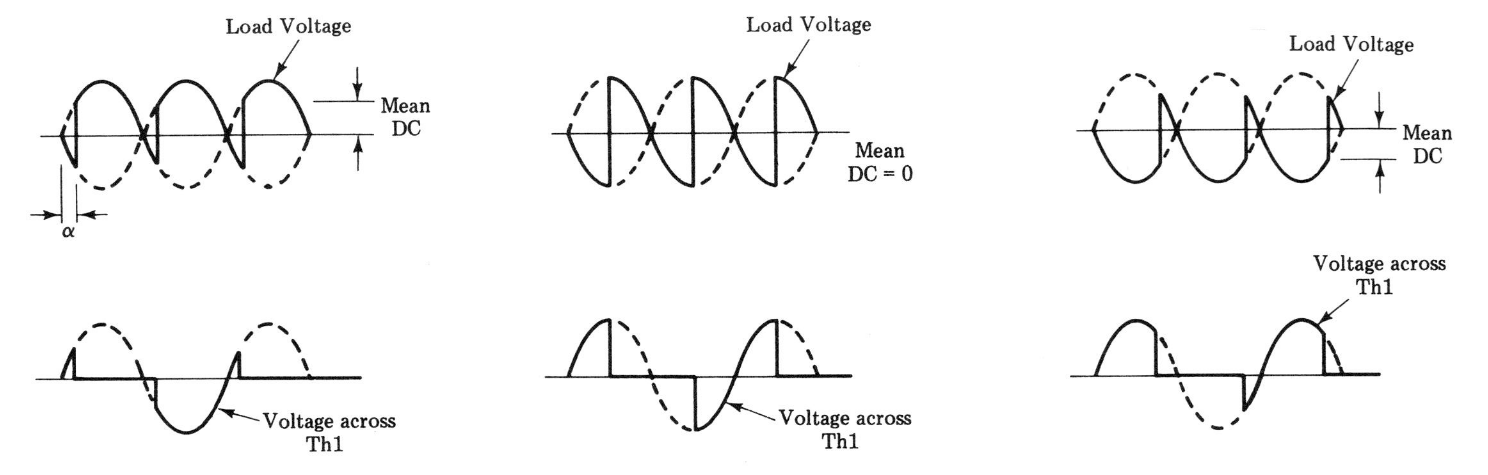

Figure 4.6. Waveforms: (a) $\alpha = 30°$; (b) $\alpha = 90°$; (c) $\alpha = 150°$.

The output waveforms with various delay angles are shown in Figure 4.6. More detailed discussion of operation in the rectification and inversion modes is given in Chapter 7.

Figure 4.6 illustrates the circuit waveforms for various delay angles. The following can be noted from these curves:

1. For a delay angle $\alpha = 30°$, more power is fed into the load than is received back into the supply.
2. The mean dc load voltage decreases as the firing angle α is increased. At $\alpha = 90°$, the voltage goes to zero.
3. At delay angles $\alpha > 90°$, the dc voltages go negative. At $\alpha = 180°$, the negative dc voltage is maximum.
4. The period for which a thyristor is reverse-biased reduces as the delay angle increases to 180°. To turn off a thyristor, it must be reverse-biased for greater than its turn-off time. Therefore, the maximum delay angle must always be less than 180°.

4.6 A Single-Phase Full-Wave Phase-Controlled Converter With Free-Wheeling Diode

A free-wheeling diode full-wave converter with inductive load is shown in Figure 4.7. The load voltage and current waveforms are also shown. The free-wheeling diode, D, is connected across the inductive load. The thyristors are triggered with a delay angle α. The α is varied to obtain a variable dc voltage at the load. It can be seen from Figure 4.7 that as the input voltage goes through zero at 180°, the load voltage cannot be negative since the free-wheeling diode, D, starts conducting and clamps the load voltage to zero volts. A constant load current is maintained by free-wheeling current through the diode. The period of thyristor and diode conduction is shown in Figure 4.7. This type of circuit cannot feed back energy to the source. The inductive energy of the load circulates current through the feedback diode which decays depending on the time constant of the load.

The steady-state duty on the free-wheeling diode is dependent on the delay angle of firing of the thyristors in the circuit.

The dc load current I_L is given by,

$$I_L = \frac{E_m}{\pi R}(1 + \cos\alpha) \tag{4.7}$$

The free-wheeling diode, D, carries the load current during the delay period α when the thyristors are not conducting. Hence the

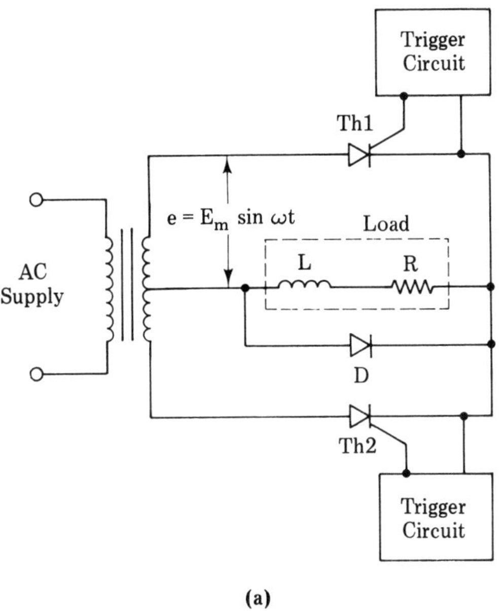

(a)

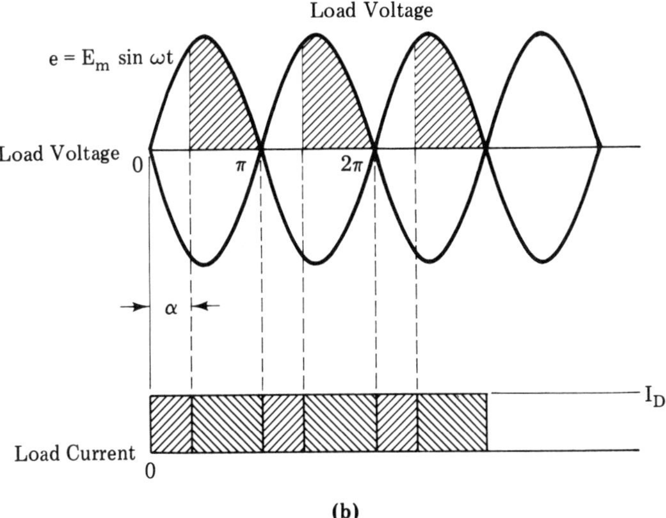

(b)

Figure 4.7. Full-wave centertap phase-controlled thyristor circuit with inductive load and free-wheeling diode: (a) circuit; (b) waveforms.

current through the diode D is given by

$$I_D = I_L \frac{\alpha}{\pi} = \frac{E_m}{\pi R}(1 + \cos\alpha)\frac{\alpha}{\pi}$$

$$= \frac{E_m}{\pi^2 R}(\alpha + \alpha\cos\alpha) \qquad (4.8)$$

4.7 The Single-Phase Full-Wave Phase-Controlled Converter Using the Bridge Principle

When an input transformer is not essential, a bridge system is often more economical. There are several variations of the bridge circuit using one thyristor, two thyristors, or four thyristors. Several variations of the single-phase bridge circuit are shown in Figure 4.8. All of the five different circuits are shown with inductive load. In the case of a resistive load, all of them operate with the same output load voltage as a function of delay angle α. The operation of each circuit is different with inductive load.

The operation of each circuit is as follows:

1. *Circuit with two thyristors and two diodes as shown in Figure 4.8(a)*

   This circuit is not acceptable with inductive load. As seen from the figure, the inductive load energy would free-wheel through D1 and Th1 or D2 and Th2. With high L/R ratio, even if gate trigger pulses are removed from the thyristors, the load current would flow through the entire negative half-cycle and hence the circuit would lose control.

2. *Bridge with two thyristors and two diodes, plus a free-wheeling diode as shown in Figure 4.8(b)*

   The operation of the circuit is similar to the one discussed above. In this case, the free-wheeling diode (being of lower impedance than a thyristor and diode in series), D3, allows the circulation of the stored energy of the load. Also during turning off of the thyristor the diode, D3, clamps the transient voltage, so the thyristors do not see any high voltage spikes.

3. *Two thyristors with two diodes connected as in Figure 4.8(c)*

   In this circuit, the energy in the load inductance circulates through diodes D1 and D2 when the thyristors are not conducting. Hence the two diodes of the bridge act as free-wheeling rectifiers. The current duty on the diodes is large in this case.

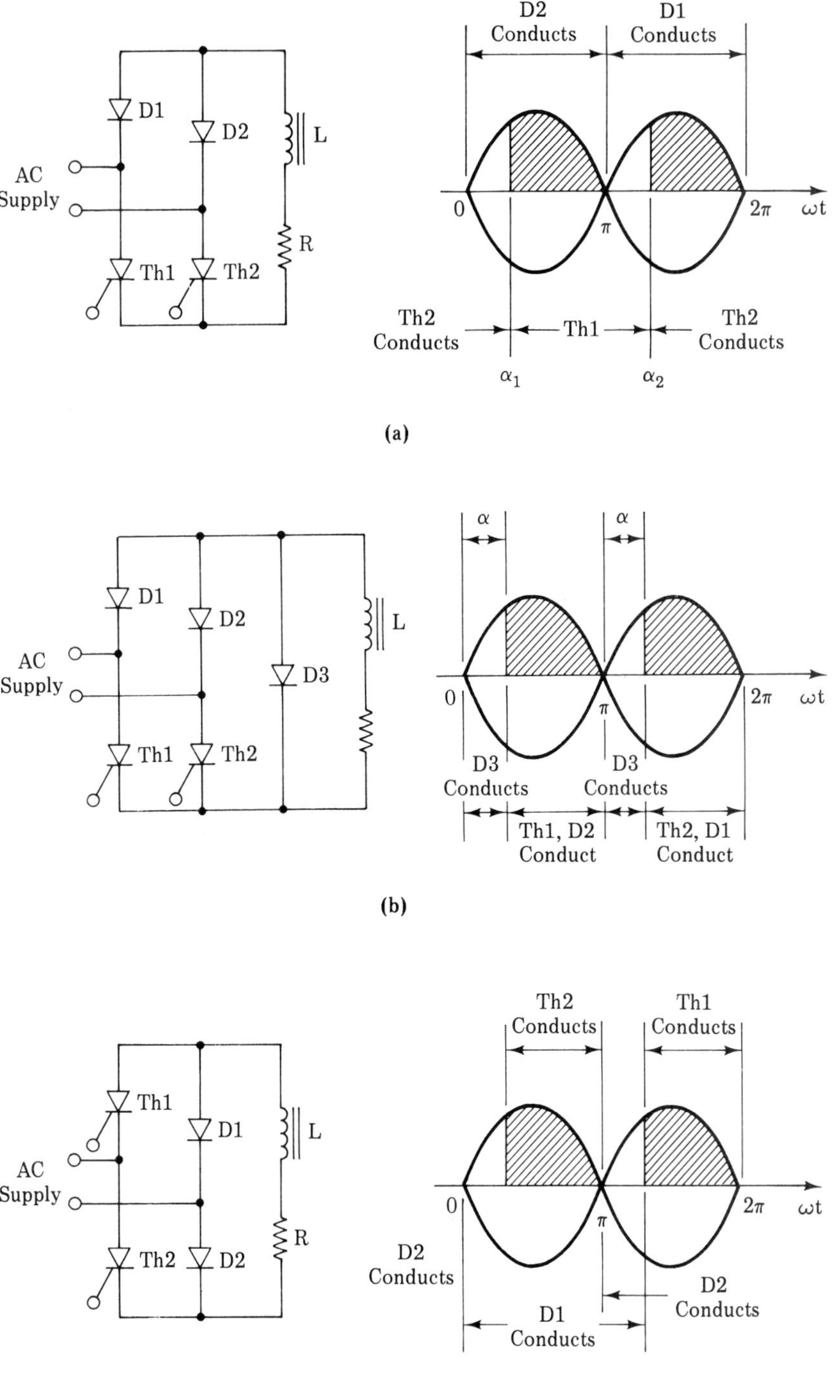

Figure 4.8. Variations of the phase-controlled bridge with inductive load (cont.).

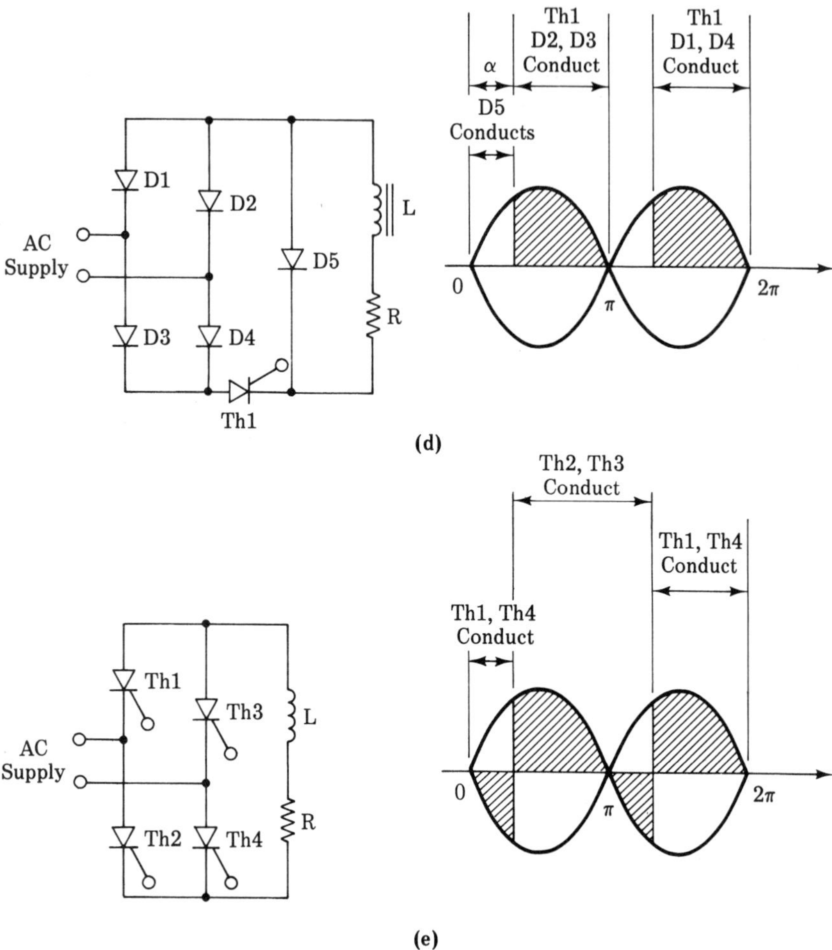

Figure 4.8. Variations of the phase-controlled bridge with inductive load (cont.).

4. *One thyristor and five diodes in a bridge connection*
   In many phase-controlled rectifier dc motor drives this configuration of the bridge is used as shown in Figure 4.8(d). Here each half of the input ac voltage is rectified by the diode bridge. The rectified full-wave voltage is then controlled by thyristor Th1. As shown in Figure 4.8(d) the output dc voltage is continuously controlled by phase-delayed triggering of Th1. The free-wheeling diode D5 allows the circulation of stored load energy during the time that Th1 is not conducting. The free-wheeling diode maintains a continuous load current even with large delay angle in triggering of Th1, required for a low output voltage. Continuous

flow of load current is essential for proper control of dc motors with minimum torque and speed ripple. The free-wheeling diode carries a large current when Th1 is operating with a large delay angle.

5. *Bridge circuit with four thyristors*

   The circuit and its associated waveforms are shown in Figure 4.8(e). None of the four previous bridge configurations is capable of feeding back load energy into the ac supply. The circuit of Figure 4.8(e) is capable of feeding back load energy into the ac source. With large inductance in the load circuit, the current conduction is continuous. The circuit is capable of operating as a rectifier or an inverter depending on the firing delay angle α of the thyristors in the bridge. This circuit acts as a rectifier for delay angle α up to 90° and for delay angles, α > 90°, the circuit operates as an inverter. This principle of operation is discussed in detail in Chapter 7.

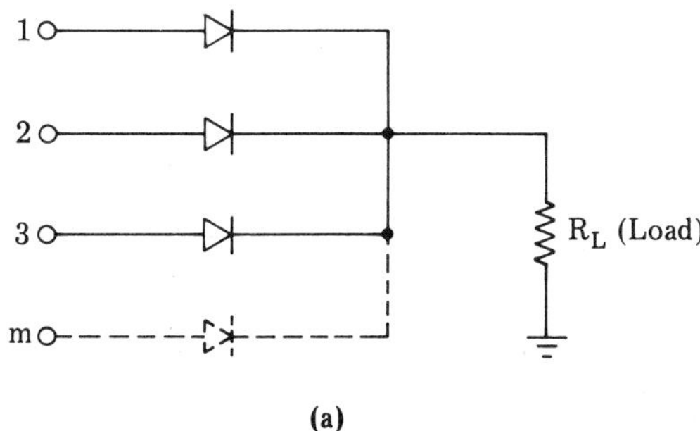

(a)

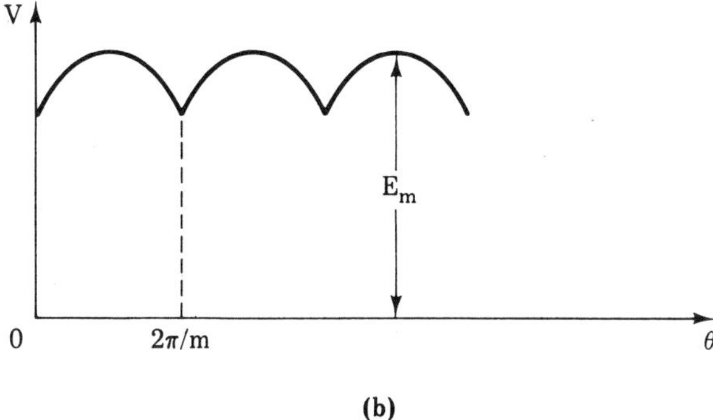

(b)

Figure 4.9. m-phase half-wave rectifier: (a) circuit diagram; (b) waveform.

4.8 m-Phase Rectifier With Resistive Load

Single-phase circuits are relatively simple in construction, but they are limited in power handling capabilities, and produce output-voltage ripples which are much greater than those from three or higher phase systems. The single-phase system can be termed two-pulse, i.e., the ratio of the fundamental dc voltage ripple frequency to that of the input ac supply is two. The greater the pulse number the lower the smoothing requirements of the circuit. A three-pulse ($m = 3$) or six-pulse ($m = 6$) system is generally used in practice.

To analyze the voltage wave of an m-phase rectifier reference may be made to Figure 4.9. The system is periodic with a period of $2\pi/m$. Over the interval $(0 < \theta < 2\pi/m)$, $f(\theta) = E_m \sin(\theta + \pi/2 - \pi/m) = E_m \cos(\pi/m - \theta)$.

Now if the system has a period $2\pi/m$, then

$$f(\theta) = a_0 + a_1 \sin m\theta + a_2 \sin 2m\theta + \cdots a_n \sin nm\theta$$
$$+ \cdots + b_1 \cos m\theta + b_2 \cos 2m\theta + \cdots$$
$$+ b_n \cos nm\theta$$

and

$$a_n = \frac{E_m}{\pi/m} \int_0^{2\pi/m} \cos(\pi/m - \theta) \sin mn\theta \, d\theta$$

$$b_n = \frac{E_m}{\pi/m} \int_0^{2\pi/m} \cos(\pi/m - \theta) \cos mn\theta \, d\theta$$

Integrating,

$$a_n = \frac{mE_m}{2\pi} \left\{ -\frac{1}{mn - 1} \cos[(mn - 1)\theta + \pi/m] \right.$$
$$\left. - \frac{1}{(mn - 1)} \cos[(mn + 1)\theta - \pi/m] \right\}_0^{2\pi/m}$$
$$= \frac{mE_m}{2\pi} \left[-\frac{2mn}{m^2n^2 - 1} + \frac{2mn}{m^2n^2 - 1} \right] \cos \pi/m = 0$$

for all values of n. Also,

$$b_n = \frac{mE_m}{2\pi} \left\{ \frac{1}{(mn - 1)} \sin[(mn - 1)\theta + \pi/m] \right.$$
$$\left. + \frac{1}{mn + 1} \sin[(mn + 1)\theta - \pi/m] \right\}_0^{2\pi/m}$$
$$= \frac{-2mE_m}{\pi(m^2n^2 - 1)} \sin \pi/m$$

The value of a_0 is given by

$$a_0 = \frac{1}{2\pi/m}\int_0^{2\pi/m} f(\theta)\, d\theta = \frac{mE_m}{2\pi}\int_0^{2\pi/m} \cos(\pi/m - \theta)\, d\theta$$

$$= \frac{mE_m}{\pi} \sin \pi/m$$

Collecting all the coefficients of Fourier analysis,

$$a_0 = E_m \cdot m/\pi \sin \pi/m \tag{4.9}$$

$$a_n = 0 \tag{4.10}$$

$$b_n = \frac{-2E_m}{m^2n^2 - 1} \cdot \frac{m}{\pi} \sin \pi/m \tag{4.11}$$

Also, $$\frac{\text{rms value of the harmonic}}{\text{mean dc value of the rectified wave}} = \frac{\sqrt{2}}{m^2n^2 - 1} \tag{4.12}$$

By means of the last four equations the ripple voltage conditions of any m-phase rectifier may be expressed, provided $m > 1$.

Three-phase rectification ($m = 3$):

$$e = \frac{E_m}{\pi}\left[\frac{3\sqrt{3}}{2} - \frac{3\sqrt{3}}{8}\cos 3\theta - \frac{3\sqrt{3}}{35}\cos 6\theta \cdots\right] \tag{4.13}$$

Six-phase rectification ($m = 6$):

$$e = \frac{E_m}{\pi}\left[3 - \frac{6}{35}\cos 6\theta - \frac{6}{143}\cos 12\theta \cdots\right] \tag{4.14}$$

The results of an m-phase rectifier may be tabulated as shown in Table 4.1. Harmonics are given in Table 4.1 with reference to supply frequency.

Table 4.1. *Harmonics in m-Phase Rectifier*

Harmonic	*Harmonic rms Value / Mean dc Value of Rectified Wave percent*			
	m			
m n	2	3	6	12
2	47.2	—	—	—
3	—	17.7	—	—
4	9.44	—	—	—
6	4.05	4.05	4.05	—
8	2.25	—	—	—
10	1.43	—	—	—
12	0.99	0.99	0.99	0.99

4.9 m-Phase Controlled Rectifier With Resistive Load

1. $\alpha < \pi/2m$

For delay angle $\alpha < \pi/2m$ the output average voltage is given by

$$E_d = \frac{m}{2\pi}\int_{-\pi/m+\alpha}^{\pi/m+\alpha} E_m \cos\theta \, d\theta = \frac{m}{2\pi}\left[E_m \sin\theta\right]_{(-\pi/3+\alpha)}^{(\pi/3+\alpha)}$$

$$= \frac{mE_m}{\pi} \sin \pi/m \cos \alpha \qquad (4.15)$$

2. $\alpha > \pi/2m$

For delay angle $\alpha > \dfrac{\pi}{2m}$ the output average voltage is given by

$$E_d = \frac{m}{2\pi}\int_{-\pi/m+\alpha}^{\pi/2} E_m \cos\theta \, d\theta = \frac{mE_m}{2\pi}\left[1 - \sin(\alpha - \pi/m)\right]$$

$$= E_{do} \frac{[1 - \sin(\alpha - \pi/3)]}{2 \sin \pi/3} \qquad (4.16)$$

where $E_{do} = \dfrac{mE_m}{\pi} \sin \pi/3$ (4.17)

4.10 Three-Phase Full-Wave Bridge Connection

Bridge connections consist of two commutating groups in series and are very economical circuits. The three-phase bridge connection is widely used for power conversion as shown in Figure 4.11.

The three-phase bridge is supplied from a three-phase system and no transformer connection is necessary. The three-phase bridge is necessary to reduce the direct voltage ripple as well as the distortion of the alternating current over the two- or three-pulse system. This is accomplished by combining two three-pulse commutating groups in a bridge connection, as illustrated by Figure 4.12.

One set of SCRs is connected to the positive output terminals and furnishes a voltage represented by the solid curve of section a, while a second group supplies the minus pole with a voltage according to the broken curve. Thus, a load connected from point a to point b is supplied with the total of the two voltages. Since the two commutating groups operate with opposite polarity, their ripple voltages are displaced, and the total voltage shows six-pulse ripple. This is the most valuable attribute of the three-phase bridge connection.

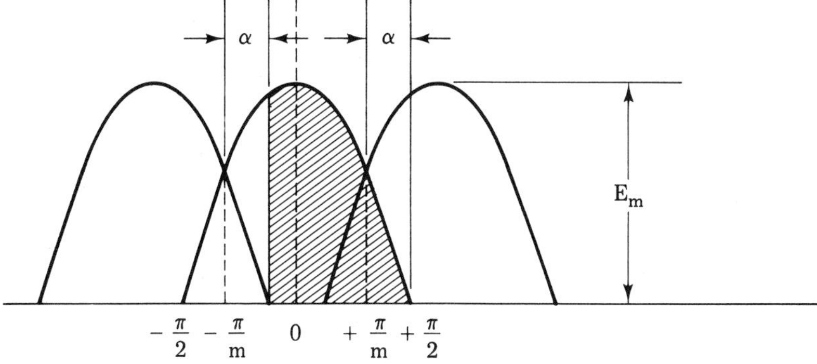

Figure 4.10. Output waveshape of an m-phase rectifier (α = delay angle).

Figure 4.12(a) shows the alternating and direct voltages with respect to the center point of the transformer, while section (c) shows the total direct voltage. The peak value of the direct voltage of each section is E_m, the peak value of the voltage a-b equals the peak value of the line-to-line voltage, $\sqrt{3}E_m$. The average value of the total direct voltage, however, is twice as high as the average value of the direct voltage of each side. The waveforms of Figure 4.12 are shown with delay angle $\alpha = 0$. The output dc voltage may be varied by delayed firing of SCRs in the bridge.

The currents in the secondary transformer windings are caused by both wyes. Each wye contributes with a 120° current pulse, and since the two current pulses are opposite in polarity, the currents in

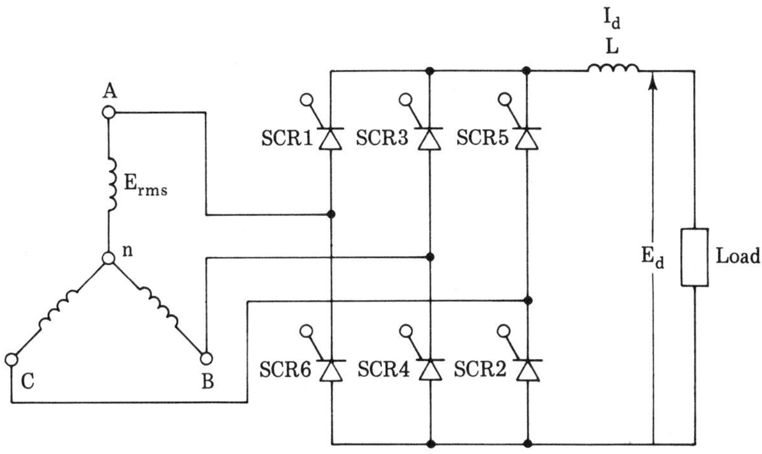

Figure 4.11. The basic three-phase bridge circuit.

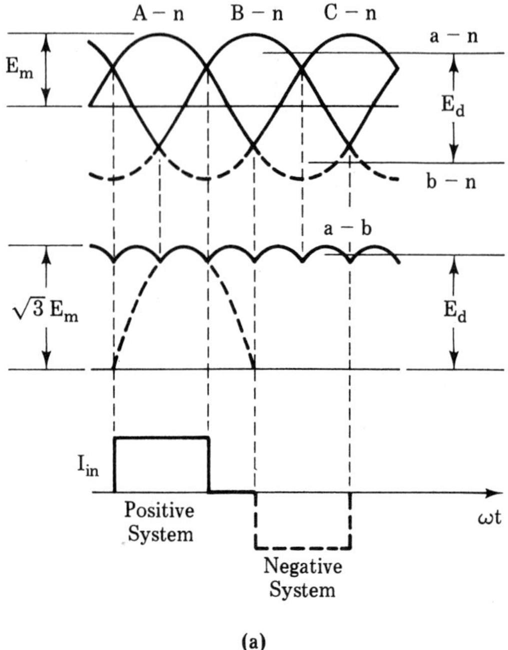

(a)

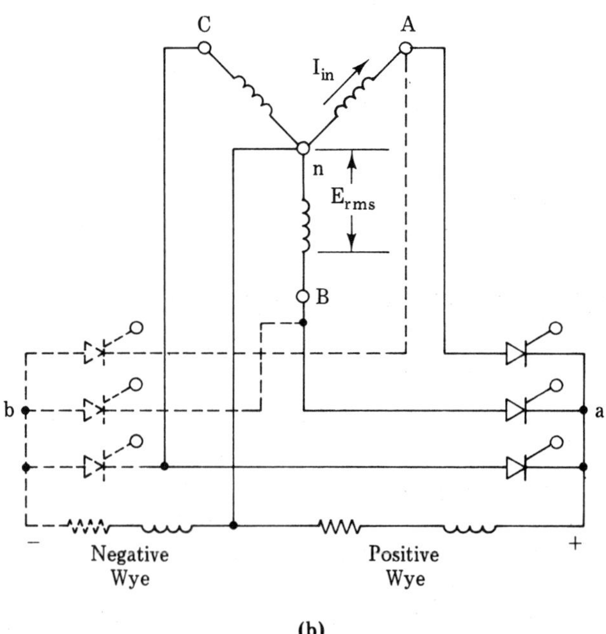

(b)

Figure 4.12. Operation of three-phase bridge circuit.

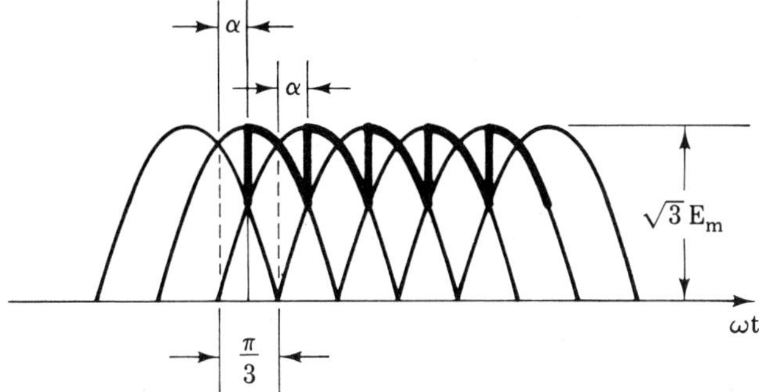

Figure 4.13. The output voltage waveform of the 3-phase full-wave bridge.

the secondary transformer windings do not have a dc component. Thus they can be balanced by the primary currents. Furthermore, without the secondary center point connection the secondary currents must total zero, which gives evidence that the primary currents must all add up to zero (ignoring the exciting currents), so that the primary winding may be in wye or delta.

4.11 Analysis of the Three-Phase Full-Wave (Bridge Connection) Circuit

The operation of the three-phase full-wave phase-controlled rectifier has been explained in the previous section. The output dc voltage of the rectifier is a function of the phase-delayed firing of the SCRs in the circuit. A fully controlled bridge circuit with six SCRs is shown in Figure 4.11. In analyzing the circuit it is assumed that the current conduction is continuous in the SCRs and the output current essentially remains constant during conduction period. The above assumption is valid in circuits with sufficient inductance.

The dc voltage at the output may be calculated from Figure 4.13 as follows:

$$E_d = \frac{1}{\pi/3}\int_{-\pi/6+\alpha}^{\pi/6+\alpha} \sqrt{3}E_m \cos \omega t \, d(\omega t) = \frac{3}{\pi}\sqrt{3}E_m \cos \alpha \tag{4.18}$$

From Eq. 4.18 it is seen that the output voltage varies as a function of delay angle, α. The output voltage is maximum when $\alpha = 0°$, and it is zero when $\alpha = 90°$. The phase-controlled circuit acts as a rectifier up to delay angle $\alpha = 90°$. With delay angles above 90° and up to 180° the output dc voltage is negative and the phase-controlled bridge is operating in the inverting mode. The output voltage is maximum negative value at $\alpha = 180°$, when $E_d = \frac{-3}{\pi}\sqrt{3}E_m$ volts.

Hence, a phase-controlled rectifier with sufficient inductance in the circuit can operate as a rectifier or as an inverter by proper choice of delay angle of firing the SCRs.

The currents in the circuit are as follows when $\alpha = 0°$:

Average value of load current $= I_d$

Average value of SCR current $= I_{d/3}$

RMS value of SCR current $= I_{d/\sqrt{3}}$

RMS value of transformer winding current $= \sqrt{\frac{2}{3}}I_d$

chapter

FIVE

AC Line Voltage Control

Phase-control techniques give the simplest ac regulating system and are the best known. However, there are other methods by which thyristor control of ac lines is possible, namely zero voltage switching, ac chopper control, and synchronous tap changing. AC line voltage controllers are employed to vary the rms value of the alternating voltage applied to a load by using thyristors as a switch.

AC phase-control methods are discussed in detail and other methods of rms voltage control are also given.

The applications of ac line voltage controllers include the following:

1. Lighting controls
2. Induction heating
3. Industrial heating
4. Transformer tap changing
5. Speed control of induction motor

5.1 AC Phase Control With Resistive Load

The basic forms of the ac phase-control circuit are given below. The power circuits and output voltage waveforms of half-wave and full-wave voltage controllers with resistive loads are shown in Figure 5.1.

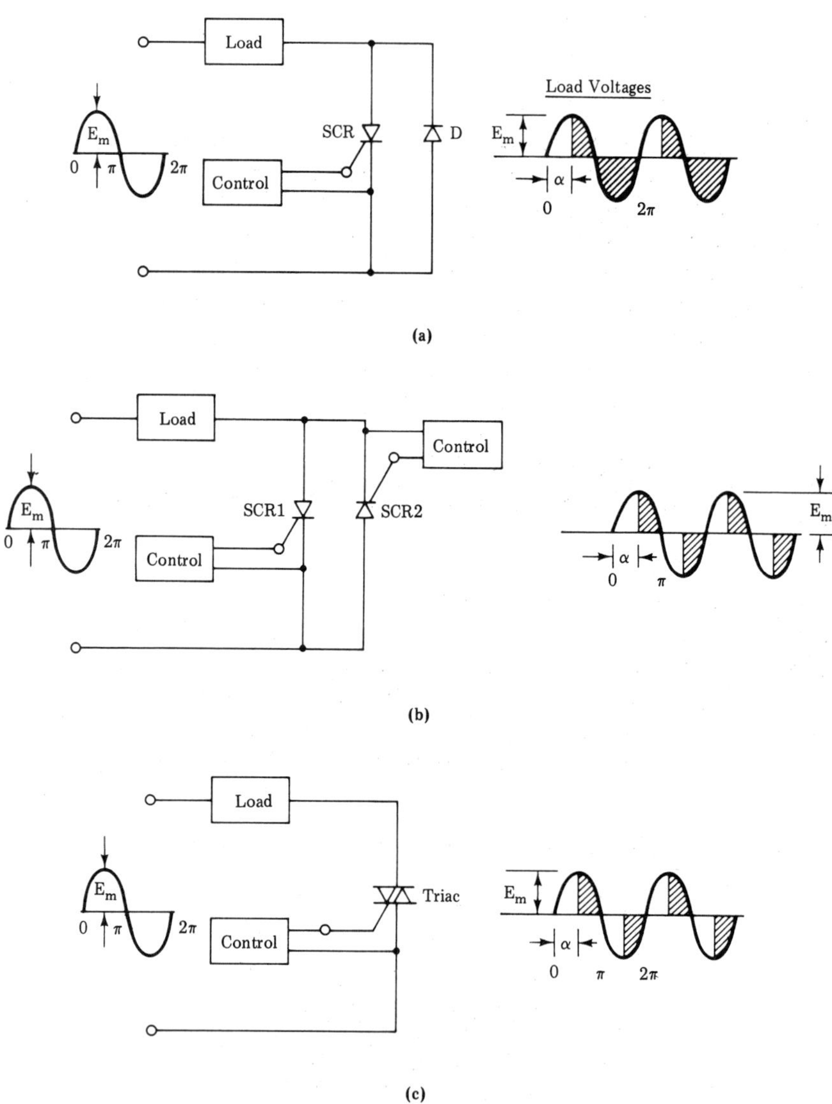

Figure 5.1. AC voltage controller circuits: (a) half-wave controller—1 SCR, 1 diode; (b) full-wave controller—two SCRs; (c) full-wave controller—triac.

The circuit in Figure 5.1(a) provides a fixed half-cycle of power and a power control range of half-power minimum and full-power maximum, but with a strong dc component. The use of back-to-back SCRs as shown in Figure 5.1(b) allows symmetrical control of ac voltage during both halves of the ac wave and eliminates any dc component.

The most simple, efficient and reliable method of controlling ac power is by using the triac, as shown in Figure 5.1(c). The triac is a bidirectional device, and a single control circuit is sufficient to control power flow in the load during both halves of the ac input wave.

5.2 Analysis of a Full-Wave Controller: Resistive Load

The single-phase full-wave controller is widely used in lighting control. The circuit of Figure 5.1(c) using the triac is a typical application for lighting control.

Assuming a pure resistive load, the waveforms at various points in the circuit are shown in Figure 5.2.

Let

E = rms input voltage of ac line
R = load resistance
α = firing delay angle
V_L = rms load voltage
I_L = rms load current

From Figure 5.2, the rms load voltage, V_L, and rms load current, I_L, are given by:

$$V_L = \left[\frac{1}{\pi}\int_{\alpha}^{\pi} (\sqrt{2}E)^2 \sin^2\theta \, d\theta\right]^{1/2}$$

$$= E\left[1 - \frac{\alpha}{\pi} + \frac{\sin 2\alpha}{2\pi}\right]^{1/2} \tag{5.1}$$

and

$$I_L = \frac{E}{R}\left[1 - \frac{\alpha}{\pi} + \frac{\sin 2\alpha}{2\pi}\right]^{1/2} \tag{5.2}$$

The rms current rating of the triac is given by

$$I_T = \frac{E}{R}\left[1 - \frac{\alpha}{\pi} + \frac{\sin 2\alpha}{2\pi}\right]^{1/2} \tag{5.3}$$

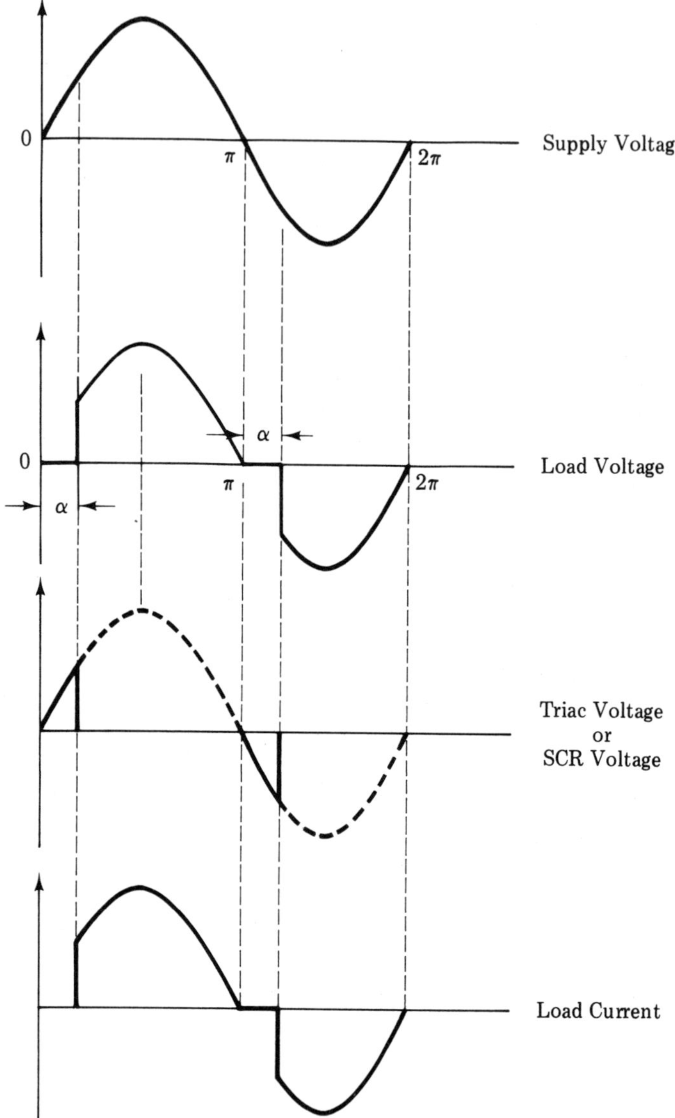

Figure 5.2. Waveforms of circuits of Figures 5.1(b) and (c).

The rms current rating of the SCR in Figure 5.1(b) is given by

$$I_{SCR} = \frac{1}{R}\left[\frac{1}{2\pi}\int_{\alpha}^{\pi} (\sqrt{2}E)^2 \sin^2\theta \, d\theta\right]^{1/2}$$

$$= \frac{E}{\sqrt{2}R}\left[1 - \frac{\alpha}{\pi} + \frac{\sin 2\alpha}{2\pi}\right]^{1/2} \tag{5.4}$$

5.3 Full-Wave Controller With Inductive Load

In the previous section the ac full-wave controller with resistive load has been discussed. In actual practice most loads have some amount of inductance. Motors, solenoids, transformers, and even some resistive heaters have inductive components as a part of their impedance. The effect of this reactance is that the rms-to-average-current ratio is lowered. In lowering this ratio, the dissipation of the device is lowered and higher average current can be safely passed through the SCR.

With an inductive load, current flow is maintained through the SCR even after the input voltage has reversed polarity and goes negative. The duration of this interval is determined by the load power factor; typical waveforms are shown in Figure 5.4.

Let us consider an interval when SCR1 is conducting. Then

$$i_{SCR1} = \frac{E_m}{Z}\left[\sin(\omega t - \phi) - \sin(\alpha - \phi)e^{(R/L)(\alpha/\omega - t)}\right] \text{A} \qquad (5.5)$$

where $\quad Z = \sqrt{R^2 + \omega^2 L^2}\ \Omega$

$$\phi = \tan^{-1}\frac{\omega L}{R}$$

The above equation could be derived by solving the differential equation of the full-wave controller with inductive load. The above equation indicates (as also can be seen from the waveforms of Figure 5.4) that the load current and voltage would be sinusoidal if delay

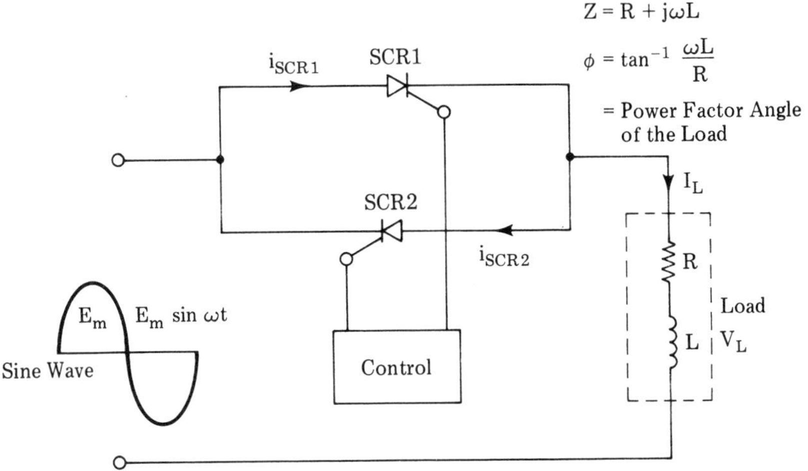

Figure 5.3. Full-wave controller with inductive load.

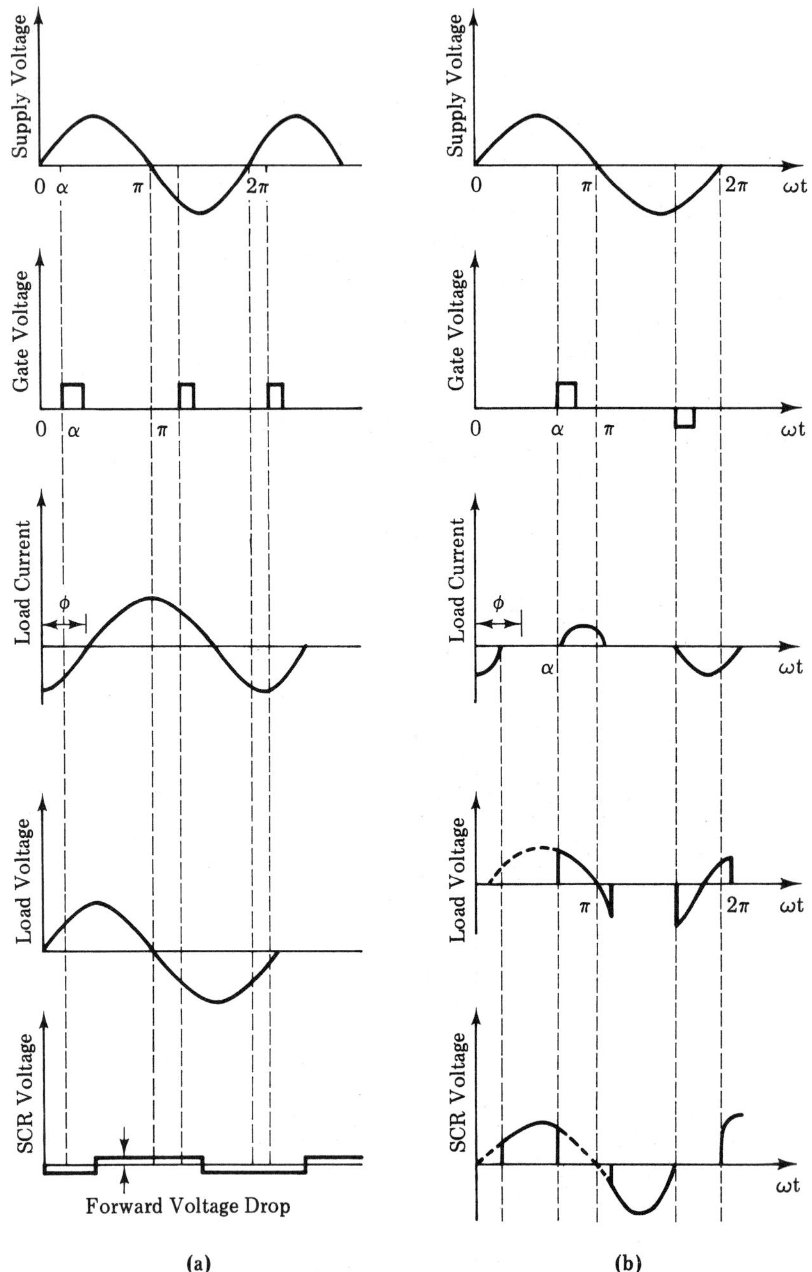

Figure 5.4. Voltage and current waveforms for a single-phase back-to-back SCR circuit with an inductive load. (a) delay angle α is less than ϕ; (b) α greater than ϕ.

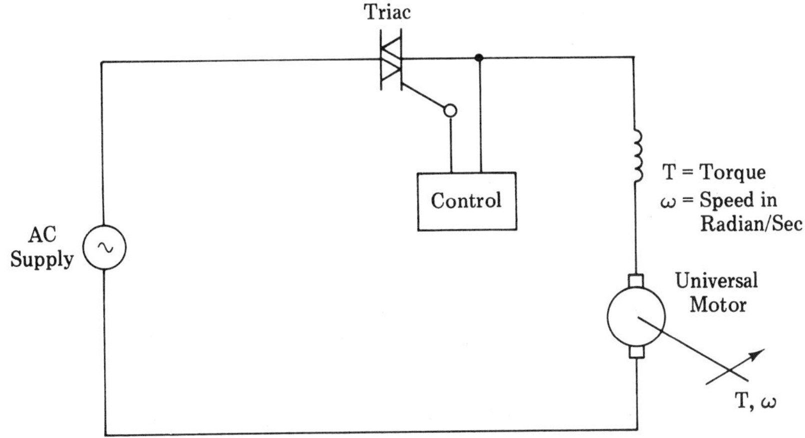

Figure 5.5. Circuit for speed control of universal motor.

angle α is less than ϕ (limit of $\alpha = \phi$). If the delay angle α is greater than ϕ, the load voltage would be discontinuous and the load current would be nonsinusoidal and discontinuous.

A triac may be used in place of the two back-to-back SCRs. These circuits can be used as power regulators for heating or lighting loads. They are also widely used for speed control of single-phase ac series or universal motors. A motor control scheme is shown in Figure 5.5.

5.4 Zero Voltage Switching

The control of rms ac voltage to the load could also be achieved by changing the on and off periods of the ac input wave to the load. There are basically two methods of control. One is referred as on-off control and the second method is called proportional control. The two methods of control are illustrated in Figure 5.6.

On-Off Control

In the on-off control as shown in Figure 5.6(a), the SCR is either fully turned on or fully turned off. The control of temperature requires a closed-loop feedback system in which the regulated temperature is sensed, compared with a reference, and a decision made whether or not power input should be changed. In the elementary system, the switch is closed at any temperature below a set point, and open at any temperature above a set point. The space temperature must fluctuate above and below the set point as shown in Figure 5.6(a) in order to

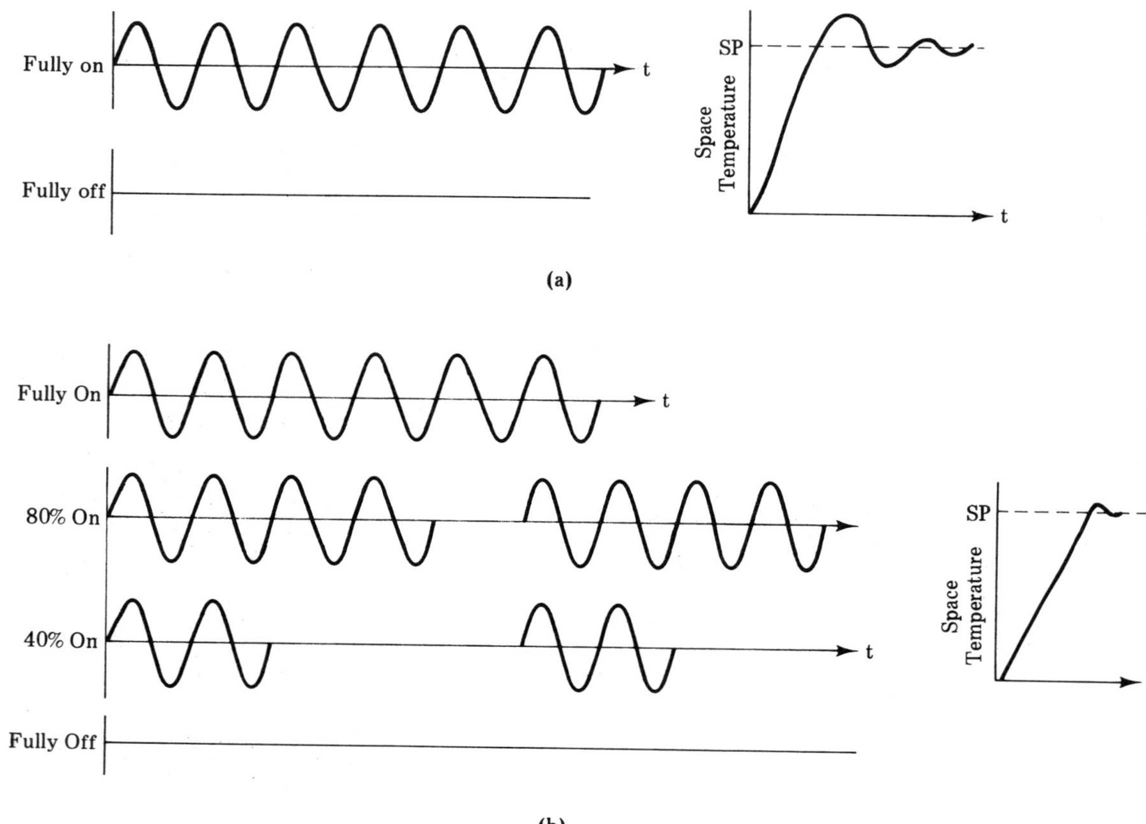

Figure 5.6. Zero voltage switching control: (a) on-off control; (b) proportional control.

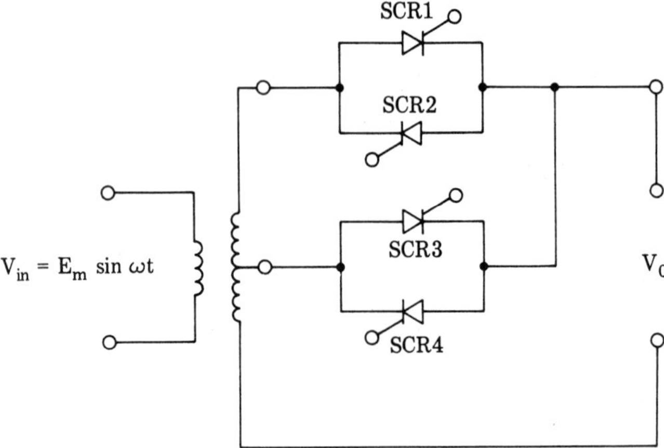

Figure 5.7. Synchronous tap changing of transformer.

establish an average input power. The fluctuation could be quite excessive in a heating system using on-off control.

Proportional Control

A good temperature control could be achieved by using a proportional control system as indicated in Figure 5.6(b). Here the average input-power is controlled in proportion to the error between space temperature and set point as in Figure 5.6(b).

5.5 Synchronous Tap Changing

If a small variation of ac output voltage is required, then the taps of a transformer would be changed by switching thyristors in the proper manner.

In the above tap changing circuit, if SCR1 and SCR2 are switched in the proper manner, then full secondary voltage of the transformer could be applied to the load, V_0. If SCR3 and SCR4 are switched, then a lower voltage from the transformer secondary is applied to the load. Thus the taps of a transformer could be changed by properly switching thyristors, and a variable ac voltage obtained at the load.

chapter

SIX

DC Choppers and DC Motor Control

Various methods are available for obtaining a variable dc voltage from a constant dc source. There is a demand for changing from one dc voltage to another. Some of the more important practical applications include power converters or armature voltage control of dc motors, converting low or high battery source voltage to levels which best match practical load requirements, and controlling dc power for a wide variety of industrial processes. The thyristor converter offers greater efficiency, faster response, lower maintenance, smaller size, and, for many applications, lower cost than motor-generator sets or gas tube approaches. In thyristor-type dc-to-dc converters, commutating circuits are required similar to those used in many inverters.

6.1 SCR Voltage Choppers

Control of a dc motor's speed by an SCR voltage chopper is required where the supply is dc (as from a battery) or the ac voltage has already been rectified to a dc voltage.

The basic circuit diagram of the dc chopper is shown in Figure

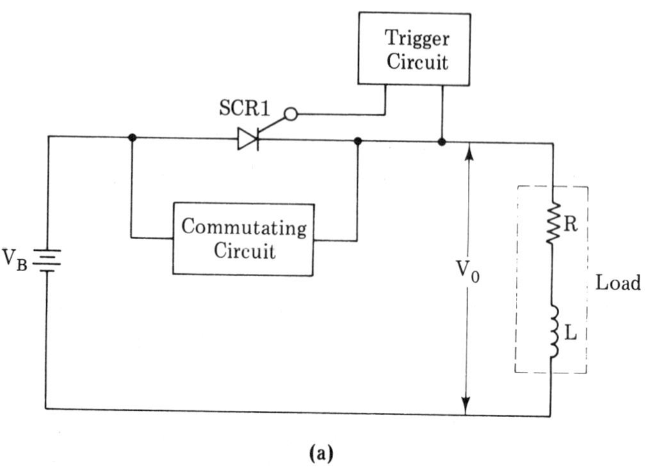

(a)

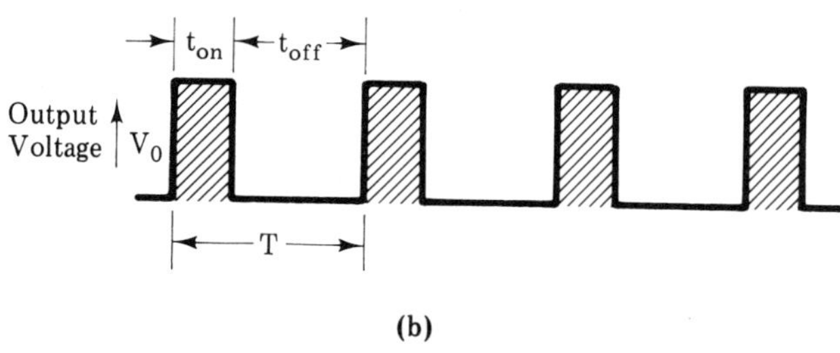

(b)

Figure 6.1. DC-DC chopper circuit: (a) basic chopper circuit; (b) output voltage waveform of the chopper.

6.1, and the principle of operation is explained below. The SCR in the circuit acts as a switch, which turns on and off the dc battery voltage to the load. The commutation circuit to turn off the SCR in the chopper is not shown. As shown in Figure 6.1(b), the battery voltage, V_B, is turned on for t_{on} seconds to the load and then turned off for t_{off} seconds. The average dc voltage across the load is given by (assuming continuous current conduction)

$$V_o = \frac{t_{on}}{t_{on} + t_{off}} \cdot V_B = \frac{t_{on}}{T} V_B \text{ volts} = \beta\, V_B \text{ volts} \tag{6.1}$$

Thus, the output dc voltage, V_o, could be varied by the following methods:

1. t_{on} may be varied, while periodic time, T, is held constant—pulse width modulation.

2. t_{on} may be kept constant while T is varied—frequency modulation.
3. Combined pulse-width and frequency modulation.

6.2 A Free-Wheeling Diode DC Motor Drive Using the Jones Chopper

The circuit under consideration is shown in Figure 6.2. This circuit is frequently referred to as a chopper circuit. The action of the SCRs is to chop the direct voltage into a series of voltage pulses. The portion of the circuit enclosed in dashed lines is the commutating circuit for SCR1. The basic idea of the circuit is that SCR1 acts as a simple switch. If the switch is closed for a long time, the motor will reach the maximum steady-state speed determined by the battery voltage, the motor, and the mechanical load characteristics. If the switch remains open, the motor will not rotate. If the switch is alternately opened and closed in a cyclic manner, the motor will rotate at some speed between zero and maximum.

The circuit is basically very efficient, since except for winding resistance and the forward conducting resistance of the SCRs and diodes, there are no dissipative elements in the circuit. The inductance, L_e (including the armature inductance of the motor), maintains the armature current through diode D1 when SCR1 is not conducting. Thus, the motor torque (proportional to the armature current) is smooth rather than pulsating. This circuit is often used with a series wound motor rather than a shunt wound motor. In a series motor, the

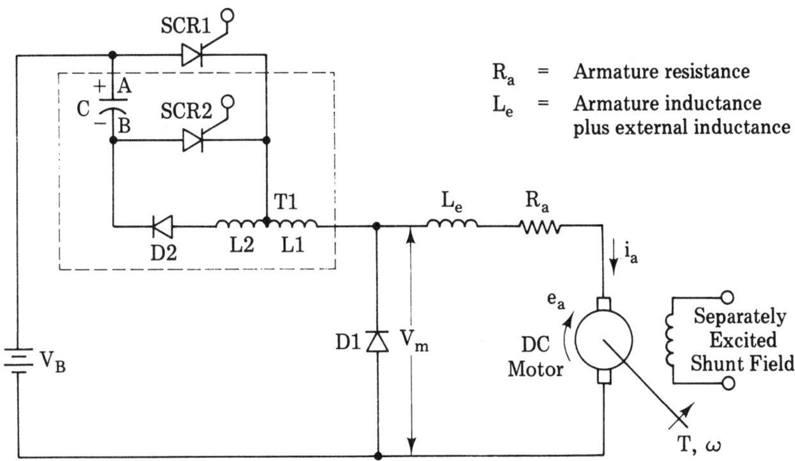

Figure 6.2. A free-wheeling diode dc motor drive.

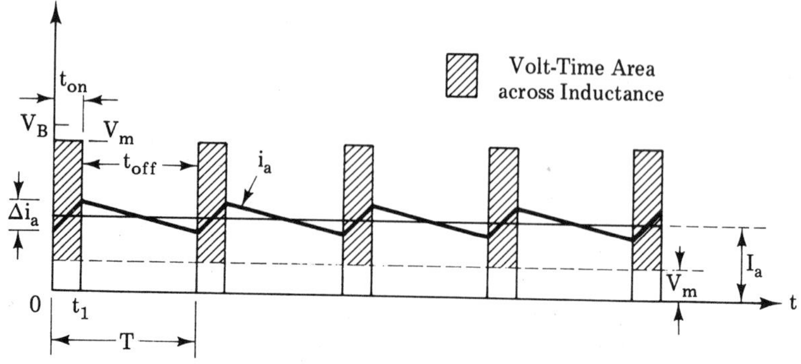

Figure 6.3. Waveforms of motor voltage and current at low speed.

series field inductance is sufficiently large that an additional external inductor may not be necessary.

The waveforms of the circuit of Figure 6.2 are shown in Figures 6.3 and 6.4. In Figure 6.3, the motor is operative at low speed. When SCR1 is turned on at $t = 0$, the armature current i_a is delivered from the battery and rises as the circuit inductance absorbs the volt-time area of the difference between V_B and the armature emf, e_a. When SCR1 is turned off after the time t_1, the armature current reduces through the free-wheeling diode D1, as the energy stored in the inductance is applied to the armature. The purpose of the free-wheeling diode D1 is to carry the inductive current when SCR1 is turned off, thus preventing high voltages from appearing across the motor. The circulating current through the free-wheeling diode also maintains a constant current through the armature and hence a constant torque. If the inductance in the armature circuit is small, the change in the armature current (Δi_a) is substantial, which would result in substantial torque ripple in the motor.

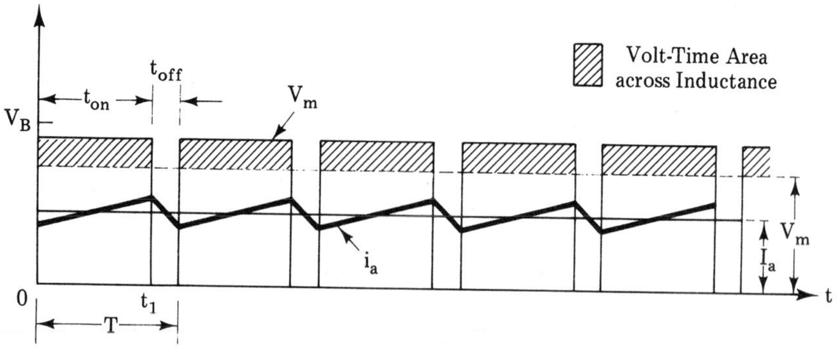

Figure 6.4. Waveforms of motor voltage and current at high speed.

Figure 6.4 shows the voltage and current waveforms of the chopper driven motor when the motor is running at high speed.

6.3 Operation of the Jones Commutation Circuit

The Jones circuit is an efficient commutating method to turn off the main SCR in the dc chopper shown in Figure 6.2. The circuit is characterized by class D commutation, that is, a charged capacitor switched by an auxiliary SCR (SCR2), and the autotransformer T1. In this circuit either t_{on} or t_{off} may be varied for voltage variation at the output.

Initially SCR1 and SCR2 are not conducting and the side B of capacitor C is charged to a negative voltage as shown in Figure 6.2. At time t1 (Figure 6.5), SCR1 is triggered, the oscillation through SCR1, L2, diode D2 and capacitor C charges up the capacitor to opposite polarity, i.e., B is positive and A is negative. The diode D2, however, prevents further oscillation of the resonating L2C curcuit. Therefore, the capacitor retains its charge until SCR2 is triggered at t3. At this point, the discharge of capacitor C reverse biases SCR1 and turns it off. The capacitor again charges up with A positive and SCR2 turns off because the current through it falls below the holding value when C is recharged. The cycle repeats itself when SCR1 is again triggered. The autotransformer action of T1 induces voltage in L2 of correct polarity for charging the commutating capacitor to a voltage higher than V_B. The bottom plate (B) of capacitor C reaches a peak value at t_5.

Since V_c (at t_5) is charged to a voltage greater than V_B, D2 is again forward biased. The capacitor now discharges to a value lower than V_B. The blocking voltages across SCR1 and SCR2 also change as seen from Figure 6.5 (t_5 to t_6). The turn-off time of SCR2 must be less than the time from t_5 to t_6. The discharging of C ceases at t_7 when current through L2 falls to zero.

6.4 Design of the Jones Commutation Circuit

The design of the commutation circuit is important in successful commutation of SCR1 in Figure 6.2. The basic design of the commutation circuit involves the proper choice of 'C' and autotransformer T1.

6.5 Commutating Capacitor

The magnitude of the commutating capacitor is dependent on the following circuit parameters:

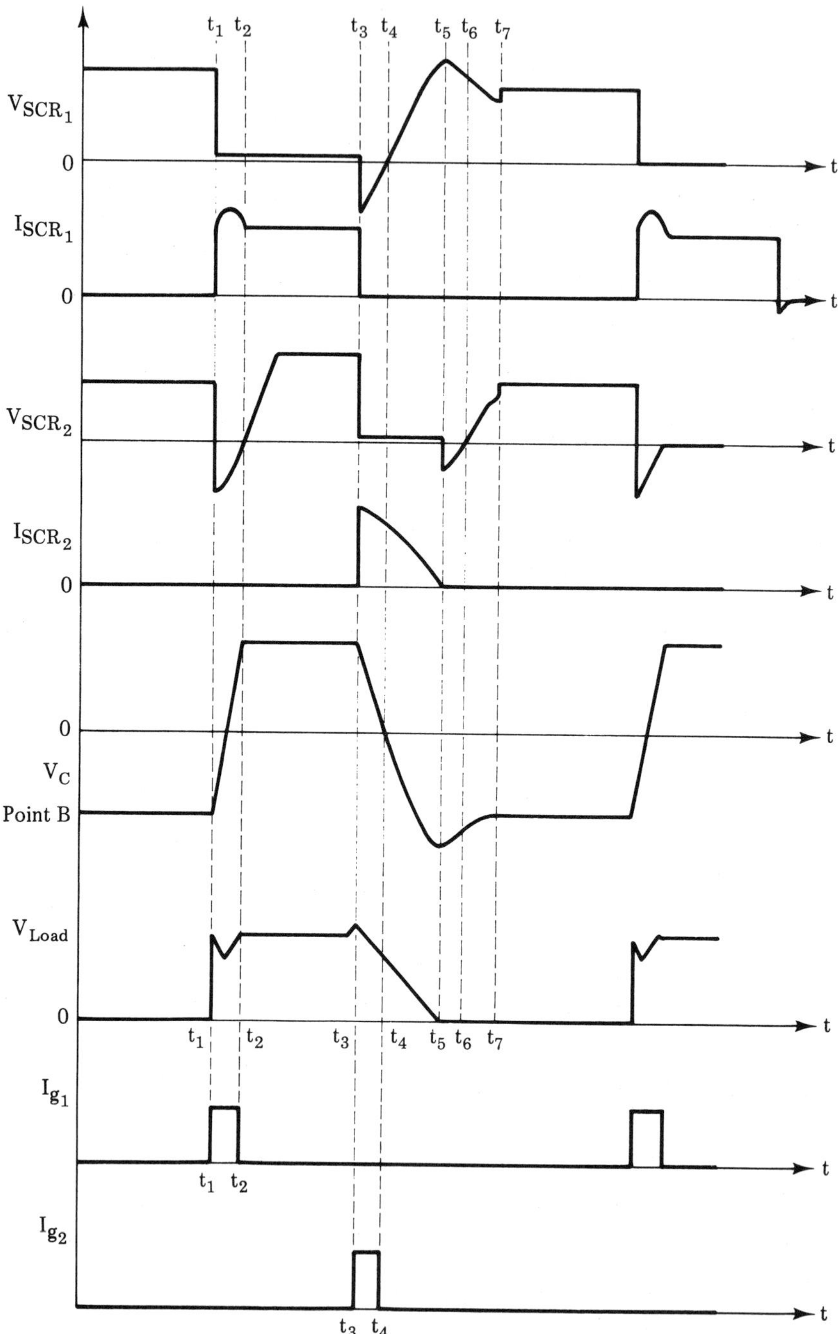

Figure 6.5. Voltage and current waveforms for the Jones chopper.

1. Maximum motor current to be commutated, I_m.
2. Turn-off time, t_o, of SCR1.
3. The battery voltage, V_B.

Initially current I_m is flowing through L1. The energy stored in inductance L1 is being transferred to capacitor C during turning off of SCR1. Hence the capacitor voltage is given by

$$V_c = I_m\sqrt{L1/C} \tag{6.2}$$

During the turn-off time, t_o, the capacitor voltage changes from V_c to 0. Hence,

$$I_m \cdot t_o = CV_c \tag{6.3}$$

Substituting the value of V_c from Eq. 6.2 into Eq. 6.3, we obtain

$$t_o = \sqrt{L1C} \tag{6.4}$$

Dividing Eq. 6.2 by V_B yields

$$\frac{V_c}{V_B} = \frac{I_m}{V_B}\sqrt{\frac{L1}{C}} \tag{6.5}$$

Defining $V_c/V_B = \alpha$, and $V_B/I_m = R_m$, then

$$\alpha = \frac{1}{R_m}\sqrt{\frac{L1}{C}} \tag{6.6}$$

Depending on the values of L1, C, and R_m, the value of α is greater than 1. The voltage across SCR1 and SCR2 is

$$V_c = \alpha V_B.$$

Thus a large value of α would require an increase in the voltage rating of the thyristor in the circuit.

EXAMPLE 6.1

An electric car is controlled by the Jones chopper as shown in Figure 6.2. The battery voltage, $V_B = 50V$; SCR turn-off time is 30 μsec, the maximum motor current is 200A. Calculate the values of commutating capacitor C and transformer inductance L1 and L2. Also, α is limited to a value of 5.

SOLUTION

From Equation 6.6, $\alpha = \frac{1}{R_m}\sqrt{\frac{L1}{C}}$, but $$R_m = \frac{V_B}{I_m} = \frac{50}{200} = 0.25\Omega$$

$$\therefore \sqrt{\frac{L1}{C}} = 5 \times .25 = 1.25 \qquad (6.7)$$

Also, from Eq. 6.4, $$t_o = \sqrt{L1C}$$

or $$30 \times 10^{-6} = \sqrt{L1C} \qquad (6.8)$$

Dividing Eq. 6.8 by Eq. 6.7,

$$C = 24\mu F$$

Also, $$L1 = 37.5\mu H$$

L2 is normally designed to be equal to L1.

EXAMPLE 6.2

An electric rapid transit system uses a dc series motor as shown in Figure 6.2. The solid-state speed controller uses a Jones chopper circuit. The motor is rated at 200 HP, 1000V, 1000 rpm, armature resistance $R_a = .05\Omega$, inductance in series with the armature is L_e (armature inductance, series field inductance and external inductance), and efficiency of the motor is 90%. The chopper frequency is 500 Hz. Find the inductance L_e required to limit the current swing in the armature under the worst condition to 5 amperes.

SOLUTION

The voltage $$V_m = \frac{t_{on}}{t_{on} + t_{off}} \cdot V_B = \beta V_B \qquad \text{(from Eq. 6.1)}$$

The voltage across the inductor L_e is given by $(V_B - \beta V_B)$.

Then, $$L_e \frac{di_a}{dt} = (V_B - \beta V_B)$$

or $$di_a = \frac{(V_B - \beta V_B)}{L_e} dt$$

From Figure 6.4, $$dt = t_{on}$$

$$\therefore di_a = \frac{V_B - \beta V_B}{L_e} t_{on} \qquad (6.9)$$

The worst condition of current swing would depend on a particular value of β. Equation 6.9 could be written as

$$di_a = \frac{V_B \quad \beta V_B}{L_e} \cdot \frac{t_{on}}{T} \cdot T$$

or
$$di_a = \frac{V_B - \beta V_B}{L_e}\beta\cdot T \tag{6.10}$$

Differentiating Eq. 6.10 with respect to β:

$$\frac{di_a}{d\beta} = \frac{V_B}{L_e}\cdot T - \frac{2\beta V_B}{L_e}\cdot T$$

For worst condition
$$\frac{di_a}{d\beta} = 0$$

or
$$(1 - 2\beta)\frac{V_B}{L_e}\cdot T = 0$$

$\therefore 2\beta = 1$ or $\beta = 0.5$ is the worst condition.

Using $\beta = 0.5$ in Eq. 6.10

$$di_a = 5 = \frac{1000 - 0.5 \times 1000}{L_e} \times .5 \times 2 \times 10^{-3}$$

$$\therefore L_e = \frac{500 \times .5 \times 2 \times 10^{-3}}{5} = 100\text{mH}$$

EXAMPLE 6.3

In the previous example find the steady-state speed and the current swing in the armature for $\beta = 0.1$ and $L_e = 100$ mH. Assume rated current in the armature for all values of β, and

$$V_B = 1000\text{V}$$

$$\beta = 0.1$$

$$\text{Power input to the motor} = P_{in}$$

$$\frac{\text{Power output}}{\text{Efficiency}} = \frac{200 \times 746}{0.9} = 165.8\text{kW}$$

Hence rated current in the armature

$$I_a = \frac{P_{in}}{V_B} = \frac{165.8 \times 10^3}{1000} = 165.8\text{A}$$

∴ armature voltage under rated torque condition is

$$e_{al} = V_B - I_aR_a = 1000 - 8.29 = 991.71\text{V}$$

For the above armature voltage the speed is 1000 rpm. For $\beta = 0.1$, voltage at the armature is given by

$$\beta \times 1000 - I_a R_a = 100 - 8.29 = 91.71\text{V}$$

$$\therefore \text{Motor speed, } N = \frac{91.71}{991.71} \times 1000 = 92.5 \text{ rpm}$$

Current swing

$$\Delta i_a = \frac{V_B - \beta V_B}{L_e} \beta \cdot T$$

$$= \frac{1000 - .1 \times 1000}{100\text{mH}} \times 0.1 \times 2 \times 10^{-3} = 1.8\text{A}$$

chapter

SEVEN

Line Commutated Inverter

The phase-controlled rectifier has been discussed in detail in Chapter 4. It was seen that the controlled rectifier enables us to rectify ac to variably dc voltage and the thyristors are naturally commutated when the ac voltage reverses polarity. In this chapter it will be shown that the controlled rectifier circuit with inductive load and phase-delayed triggering beyond 90° acts as an inverter. The basic principle of the inverter mode of operation is the same for a single-phase full-wave circuit or the three-phase full-wave circuit. Naturally the ripple frequency would be higher and the ripple magnitude would be lower for the three-phase case. Line commutated converters or inverters are used in regenerative dc/ac drives and in HVDC (high voltage direct current) transmission.

7.1 A Single-Phase Full-Wave Circuit Operating as a Controlled Rectifier and as an Inverter

A single-phase full-wave converter is shown in Figure 7.1, and the waveforms with three different phase-delay angles are shown in Fig-

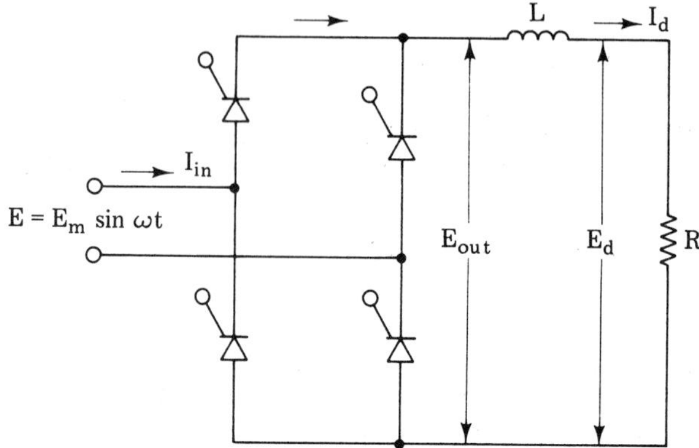

Figure 7.1. A single-phase full-wave ac/dc converter.

ure 7.2. It should be noted that since the SCRs can only conduct in one direction, the current must always flow out of the converter, and inversion may take place only if a dc source is provided and the delay angle is extended beyond 90° to produce a negative dc voltage on the output of the converter.

The following assumptions are made in the analysis of the line commutated converter:

1. The dc current is assumed to be constant over each cycle.
2. The current is continuous at all times between the ac and dc terminals.
3. If the triggering of one set of thyristors is delayed, the other set must continue conduction in the meantime.
4. When the ac source reverses polarity, the forward bias on the thyristors to maintain conduction will be supplied by the inductance.

The average voltage, E_d, may be evaluated as follows from Figure 7.2:

$$E_d = \frac{1}{\pi} \int_{\alpha}^{\pi+\alpha} E_m \sin \omega t \, d(\omega t) = \frac{2E_m}{\pi} \cos \alpha \qquad (7.1)$$

From Eq. 7.1, the average dc voltage, E_d, may be evaluated as shown in Figure 7.2. It is seen that as delay angle α varies from 45° to 90° to 135°, the average dc output voltage changes from positive to zero and to a negative value. The ac source current waveform remains a square wave of amplitude I_d, but is displaced from the voltage wave by an angle $\phi = \alpha$. Therefore the converter has an increasingly lagging power factor at the ac terminals as the output dc voltage is reduced.

Since the dc current is constant, the instantaneous power dissipated in the load resistor is the same throughout each half cycle. However, the power interchange between other parts of the circuit changes continually. When E_{out} exceeds E_d, the voltage across the inductance is positive, and both the inductance and the load resistor are absorbing power from the ac source. When E_{out} drops below E_d, the inductance starts to supply part of the power to the resistor. Then, when the source voltage reverses polarity, the inductance supplies instantaneous power to the load resistor and to the ac source. The net power flow over one cycle is from the ac source to the load resistance. It should be noted that since the SCRs can only conduct in one direction, the current must always flow out of the converter, and inversion may take place only if a dc source is provided and the delay angle is extended beyond 90° to produce a negative dc voltage at the output of the converter.

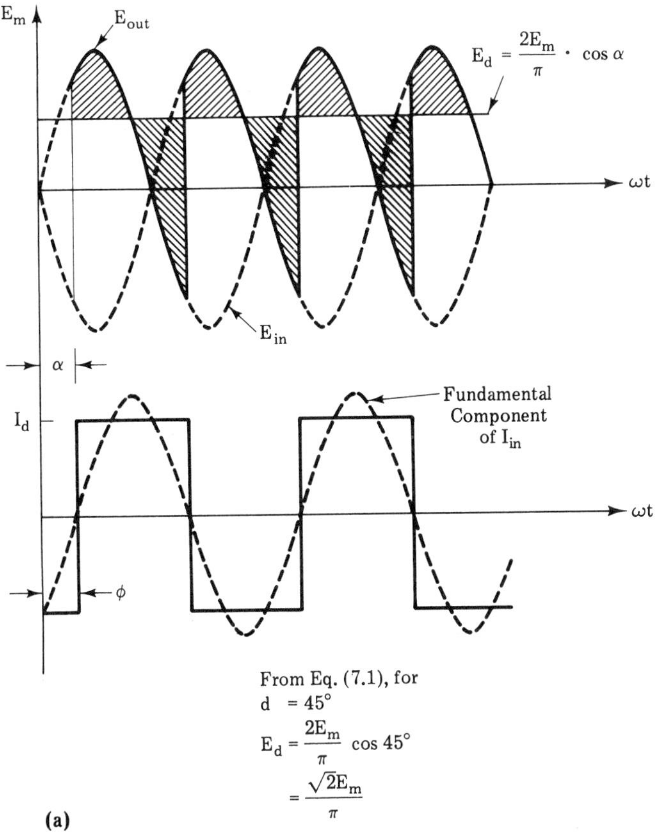

Figure 7.2. The waveforms of Figure 7.1 for various delay angles: (a) α = 45°. (*Continued*)

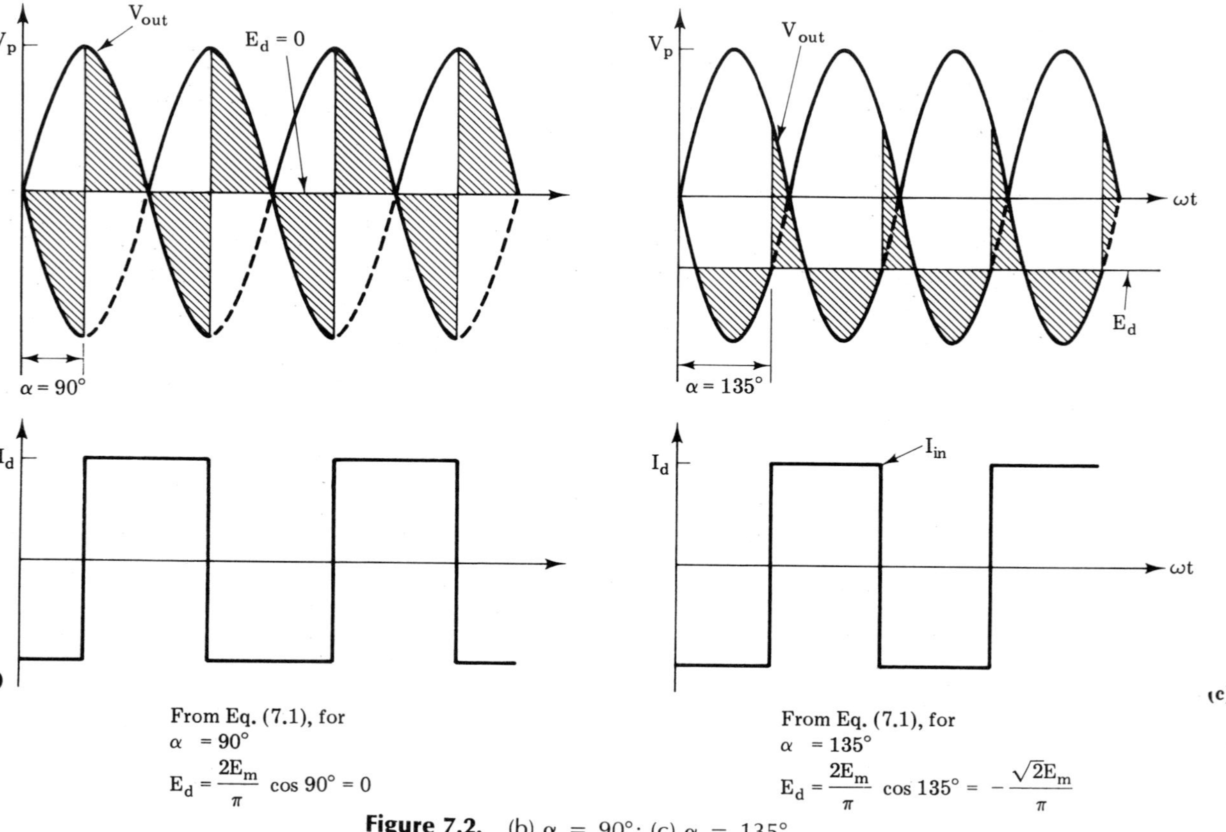

Figure 7.2. (b) $\alpha = 90°$; (c) $\alpha = 135°$.

The single-phase full-wave converter may be used in regenerative dc drive systems.

7.2 Three-Phase Full-Wave Bridge Circuit as a Converter

A converter may transfer power flow from ac to dc, in which case it operates as a rectifier, or it may transfer power from dc to ac, in which case it operates as an inverter. The same converter with full delay angle control can operate either as a rectifier or as an inverter, and the mode of operation is simply a matter of the firing angle control. The operation of the three-phase bridge circuit as a rectifier has been discussed in detail in Chapter 4. In the following section the operation of the converter in the rectifying and inverting mode is explained.

7.3 Three-Phase Full-Wave Bridge Circuit as a Phase-Controlled Rectifier and a Line Commutated Inverter

The circuit shown in Figure 7.3 with variable dc voltage at the output and sufficient inductance in the circuit acts as a rectifier or inverter, depending on the delay of triggering of the SCRs in the bridge. The waveforms of the circuit are shown in Figure 7.4. It is clearly seen from the waveforms that the circuit changes from rectification to inversion mode at $\alpha = 90°$.

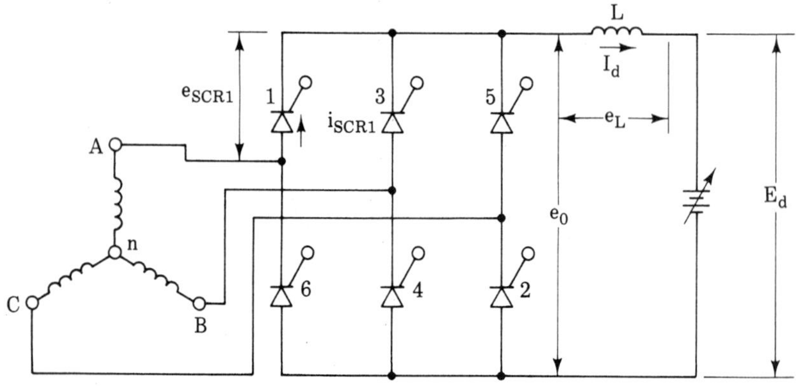

Figure 7.3. Three-phase full-wave bridge with variable dc source at the output.

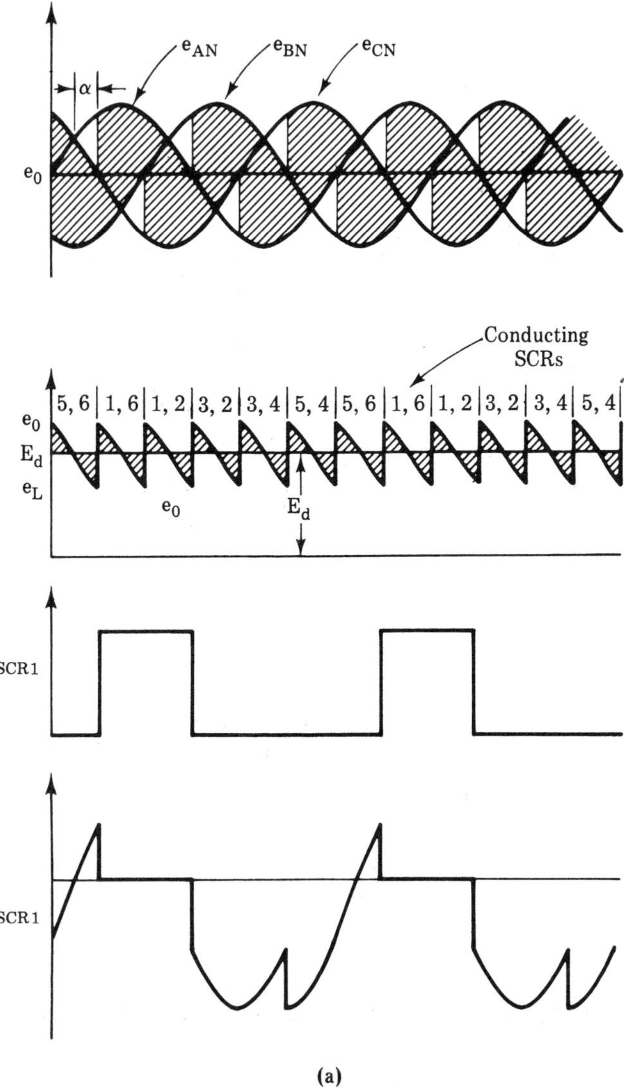

Figure 7.4. Waveforms of three-phase full-wave bridge with various delay angles: (a) $\alpha = 30°$. *(Continued)*

7.4 HVDC Transmission

The three-phase full-wave bidirectional converter may be used in dc motor drives with regenerative capability. The same converter is also used in the high-voltage dc transmission system. In this section the basics of HVDC transmission are discussed.

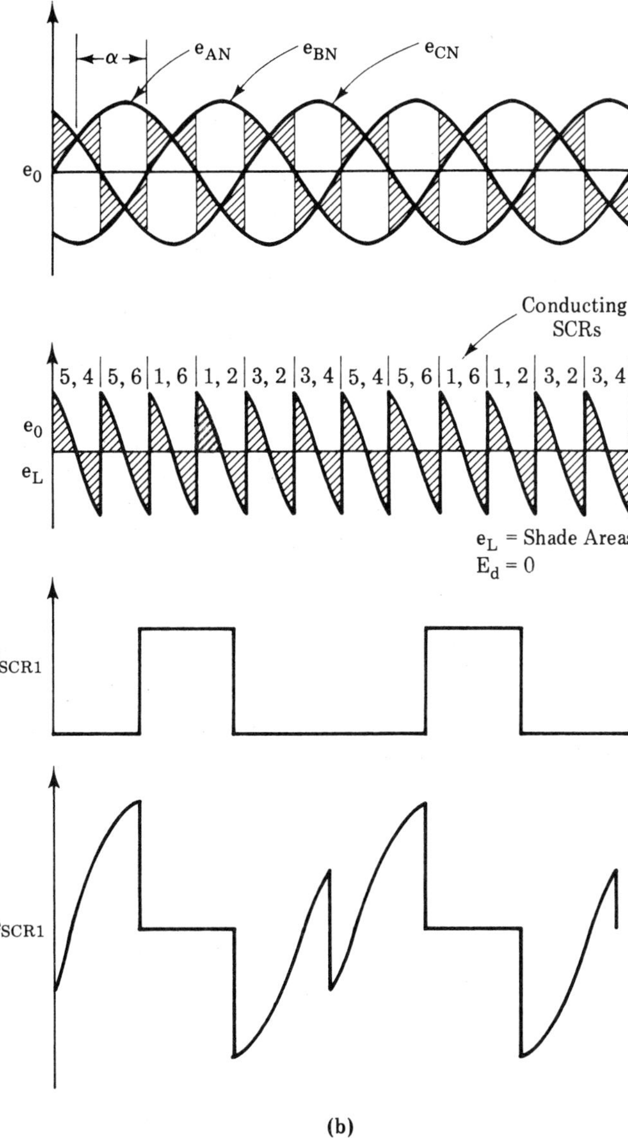

(b)

Figure 7.4. (b) $\alpha = 90°$. *(Continued)*

A high-voltage direct-current power transmission system is a link connecting two or more ac stations through converters and dc lines. A simple diagram of a two-terminal dc link is shown in Figure 7.5.

A converter may operate as a rectifier, in which case power is transferred from the ac to the dc side. Also power may be transferred from the dc to the ac side, in which case the converter acts as an

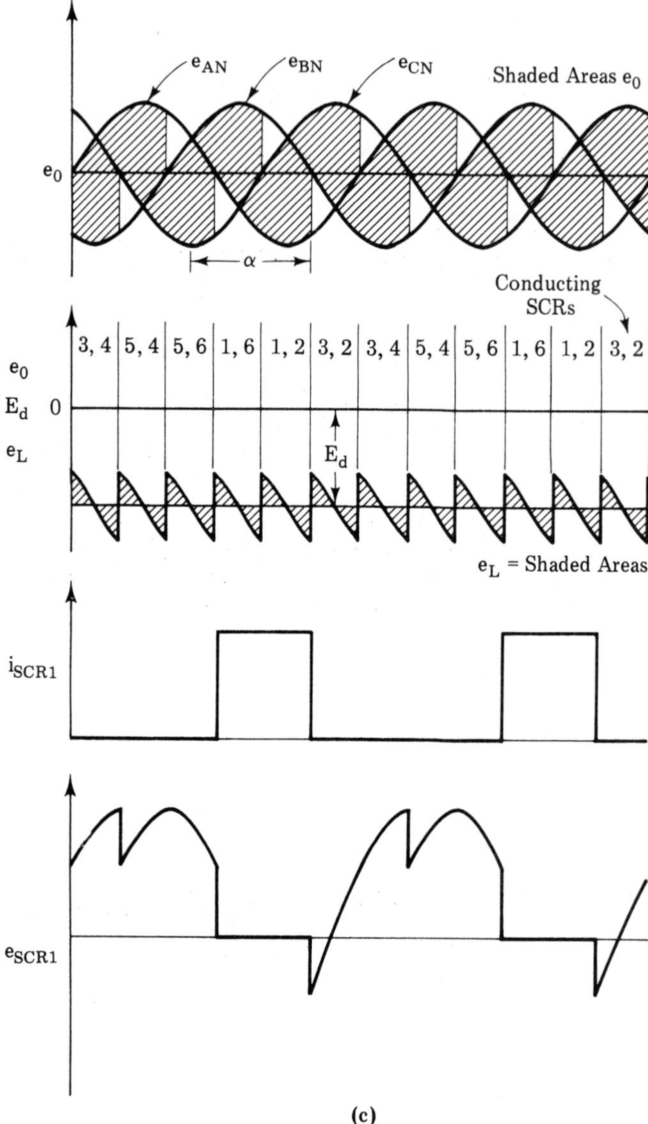

Figure 7.4. (c) $\alpha = 150°$.

inverter. A converter acts as a rectifier or an inverter, depending on the delay angle control of the thyristors (SCRs) in the converter.

There are several types of direct-current links possible in HVDC transmission. Two types of dc links are explained below:

1. The monopolar link. This has one conductor, usually of negative polarity, and ground or sea returns as shown in Figure 7.6.

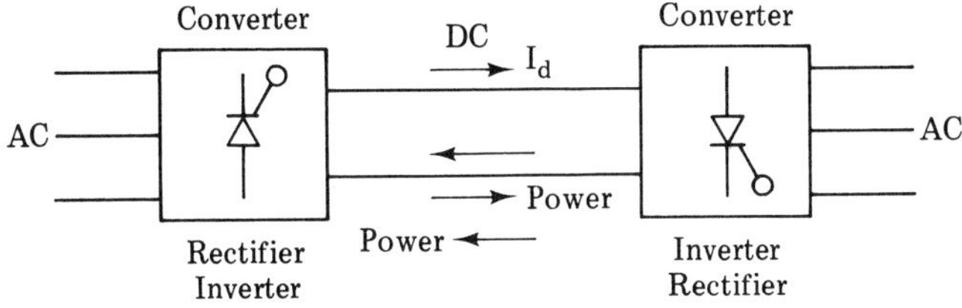

Figure 7.5. A two-terminal dc transmission.

2. The bipolar link. This system has one positive and one negative conductor. There are two converters in series at each terminal. The voltage ratings of the converters are equal and the neutral point of each may be grounded. When both neutrals are grounded, each pole can operate independently. Normally both poles operate with equal current, and hence there is no ground current. One conductor with ground return can carry half the power.

The rated voltage of a bipolar link is expressed as ±400 KV, i.e., each pole is 400 KV with respect to ground.

7.5 Pacific Northwest-Pacific Southwest Interties

The Pacific NW/SW Intertie is connecting Celilo, Oregon with Sylmar, near Los Angeles in California. The above mentioned dc transmission is a ±400 KV bipolar system. The transmission distance from Celilo to Sylmar converter stations is about 900 miles. Mercury arc rectifiers have been used for the converters. In the solid-state converter, the

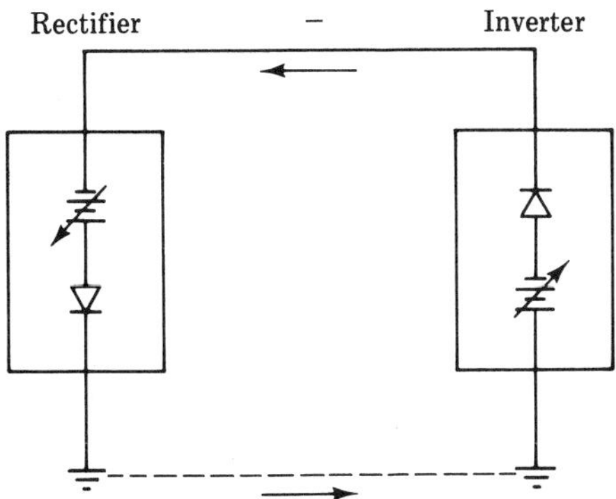

Figure 7.6. Monopolar dc link.

basic unit is the SCR. The converter station voltage and current ratings are high, and hence series and parallel connected SCRs have to be used. The first solid-state converter scheme has been used for the Eel River Project in New Brunswick, supplied by General Electric Company.

7.6 Principle of Operation of the Converter

The equivalent circuits for the Pacific NW/SW Intertie are shown in Figures 7.7 and 7.8. By proper delay angle control of SCRs, the dc output voltage of the converter could be varied from zero to positive maximum and then to negative maximum. The converter can conduct in one direction only. The voltage and current parameters are shown in the equivalent circuit (Figures 7.7, 7.8) corresponding to maximum power flow in Pacific NW/SW Intertie. The resistance of the dc transmission line between the two converter stations is 19Ω. The maximum current capability of the line is 1800A. When the dc output voltage at the Celilo converter is +400 KV, the Sylmar converter output has to be adjusted to 365.8 KV, for maximum power flow through the line. In Figure 7.7, the power flow from Celilo is 720 MW and the power received at Sylmar is 658.44 MW. The power loss in the transmission line is 61.56 MW.

In Figure 7.8, the output voltage of the Sylmar converter is −400 KV and that of the Celilo converter is −365.8 KV. The direction of current flow is the same as in the previous case, but now the Sylmar converter is acting as a rectifier delivering 720 MW and the Celilo converter is acting as an inverter receiving 658.44 MW.

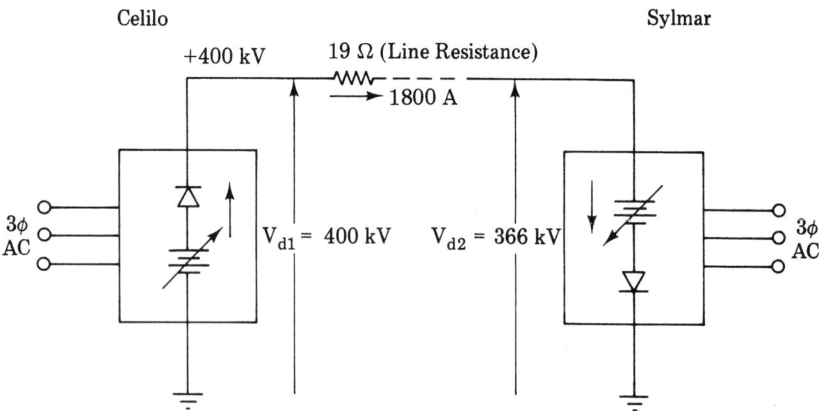

Figure 7.7. Equivalent circuit of Pacific NW/SW Intertie (power flow from Celilo to Sylmar).

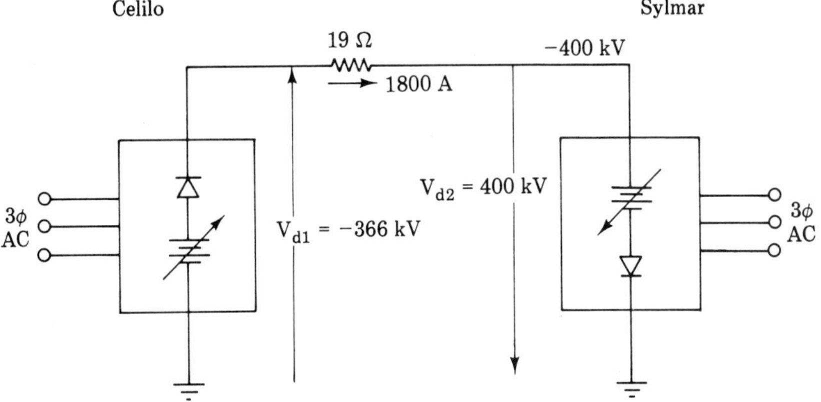

Figure 7.8. Equivalent circuit of Pacific NW/SW Intertie (power flow from Sylmar to Celilo).

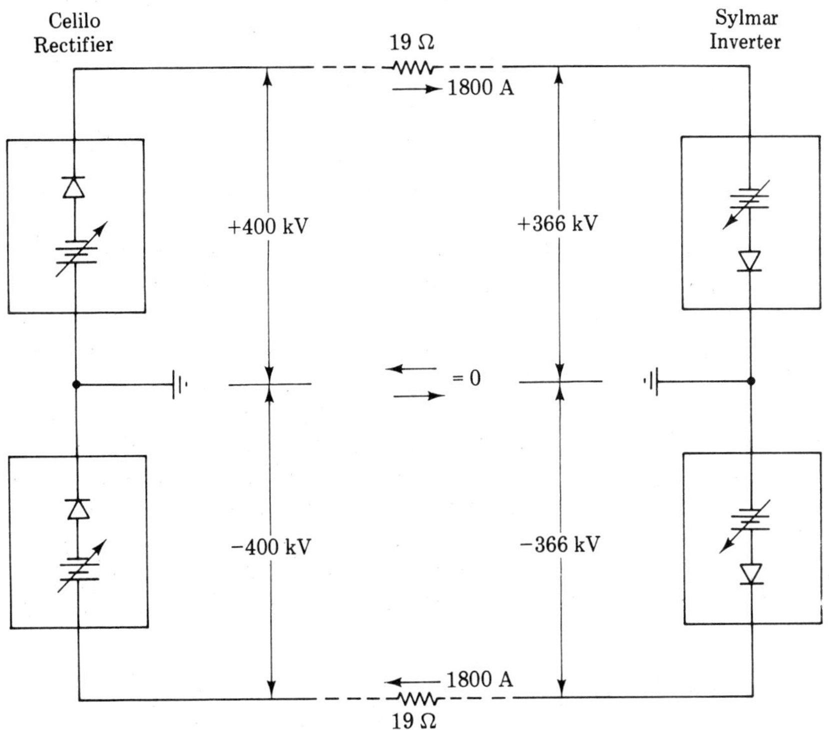

Figure 7.9. Bipolar system of Pacific NW/SW Intertie.

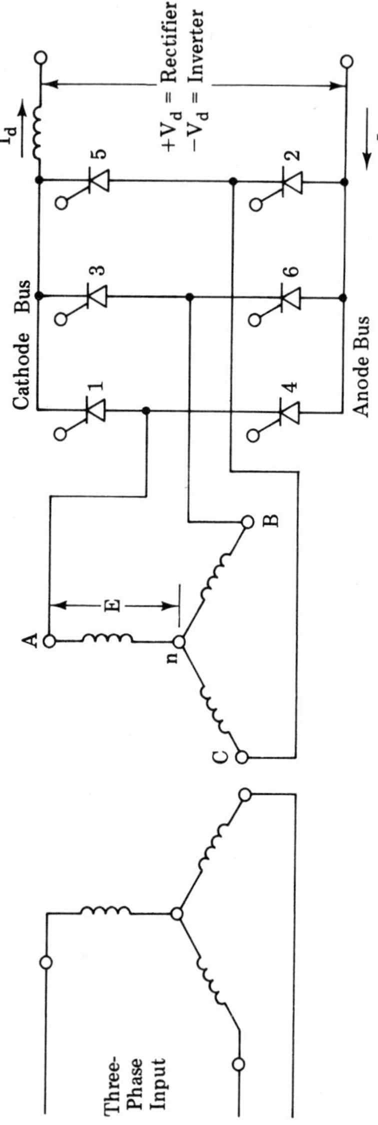

Figure 7.10. Three-phase bridge converter.

A bipolar system is shown in Figure 7.9. The net ground current is zero when the two poles are carrying identical current. The total power of the bipolar system is 1440 MW. The power flow in the line could be controlled from minimum to maximum by proper delay angle control of the SCRs in the converter. The output dc voltage of the converter has considerable ripple. The ripple is smoothed by dc reactors and the output dc current is assumed to be smooth.

The converter uses the three-phase bridge connection as shown

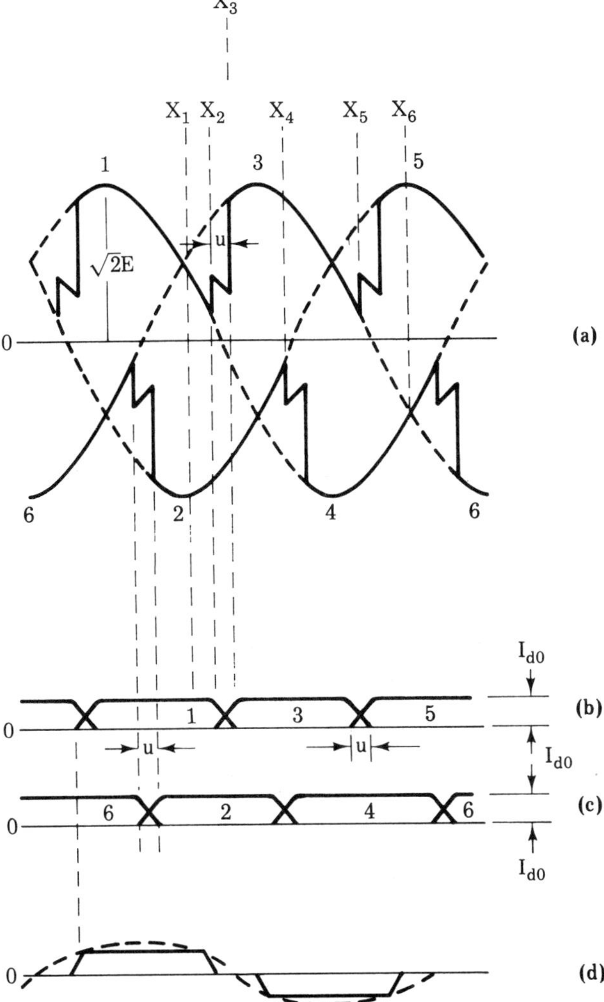

Figure 7.11. Operation of a bridge as a rectifier (numbers correspond to SCRs in Figure 7.10).

in Figure 7.10. This has been accepted as the best connection for HVDC converters. Total power capacity is met by series and/or parallel connection of groups. The operation of the bridge as a rectifier and inverter has been discussed in Chapters 4 and 7. The operation of the three-phase bridge as a rectifier and as an inverter is again described in Figures 7.10 to 7.13.

The operation of the circuit is explained through Figure 7.11. At the instant X1, SCR1 is conducting. At this point voltage of phase B becomes more positive with respect to phase A. The anode of SCR3 is now more positive than the cathode. Hence SCR3 may now be fired if a trigger pulse is applied to its gate. In the figure, SCR3 is fired with a delay angle α. Up to this time SCR1 continues to conduct and the output voltage falls. The firing of SCR3 could be delayed up to point X6 ($\alpha = 180°$), beyond which its anode becomes more negative.

When SCR3 is triggered on, the commutation of SCR1 starts. During commutation of SCR1, the current from SCR1 is transferred to SCR3 as shown in Figure 7.12. The time taken for the transfer is called overlap angle or commutation angle (angle u). If commutation is successful the transfer of current from SCR1 to SCR3 is complete at X3. The output voltage now follows the phase B voltage. SCR5 is fired at X5 and similarly current is transferred from SCR3 to SCR5, and the output voltage follows phase C, Thyristors 2, 4, and 6 go through a similar process on the opposite half cycles.

For the whole circuit, therefore, the output voltage is shown in Figure 7.11 by the voltage between the two thick lines. The sequence of firing of the thyristors and the current through them are shown in Figures 7.11(a) and 7.11(b).

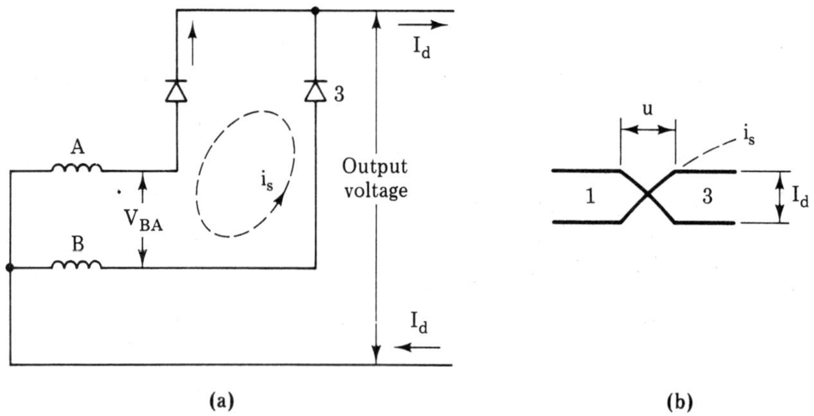

Figure 7.12. Illustration of overlap angle: (a) output voltage; (b) current in SCRs 1,3.

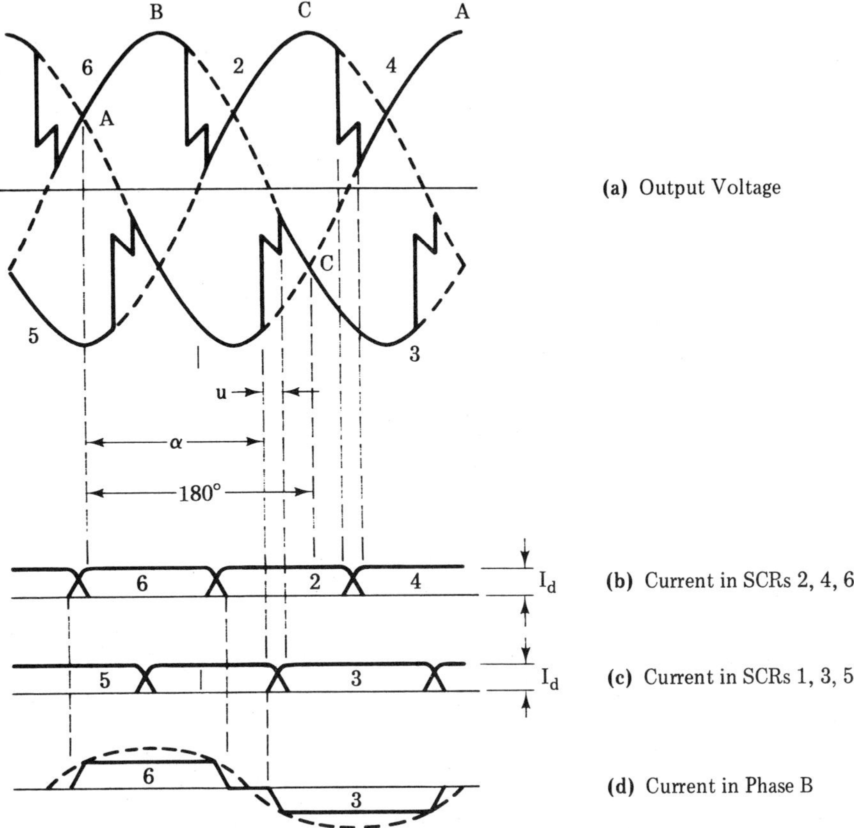

Figure 7.13. Operation of a bridge as an inverter voltage and current waveforms: (a) output voltage, 5; (b) current in SCRs, 2,4,6; (c) current in SCRs 1,3,5; (d) current in phase B.

The average output voltage is given by:

$$V_d = \frac{3\sqrt{2}}{2\pi} E[\cos \alpha + \cos(\alpha + u)] \tag{7.2}$$

If overlap angle u is neglected, then

$$V_d = \frac{3\sqrt{2}}{\pi} E \cos \alpha = V_{d_o} \cos \alpha \tag{7.3}$$

where V_{d_0} is the maximum no-load output when $\alpha = 0$.

The current waveform in phase A is shown in Figure 7.11(d) and the dotted curve is the fundamental component.

Also,

$$I_{ac} = \frac{\sqrt{6}}{\pi} I_d \tag{7.4}$$

Voltage and current waveforms when the bridge is operating as an inverter are shown in Figure 7.13. Thyristors 1, 3, and 5 conduct during the positive half of the voltage waveforms, whereas thyristors 2, 4, and 6 conduct during the negative half.

chapter

EIGHT

DC Link Inverter

Inverters are used to transfer energy from a dc source to an ac load of arbitrary frequency and phase. More specifically, they are used typically in drive systems to provide power for adjustable frequency ac motors, to regenerate energy back to the ac line from decelerating dc motors, and to pump rotor-circuit power back to the ac line from wound rotor induction motors. In non-motor systems, they are used to supply uninterruptible ac power to computers and to convert energy between ac and dc form at the terminals of high voltage dc power transmission systems.

8.1 A Single-Phase Parallel Capacitor Commutated Inverter

A thyristor inverter is composed of two parts:

1. The high power inverting elements.
2. The low power oscillator.

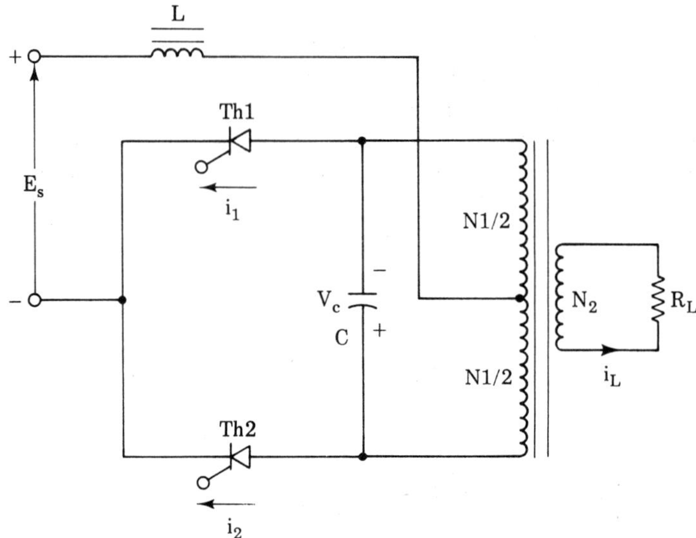

Figure 8.1. Single-phase parallel capacitor commutated inverter.

The low power oscillator is necessary to periodically trigger the thyristor or any other semiconductor switches in the inverting elements. The basic circuit is shown in Figure 8.1, and its operation is explained below, using thyristors as switching elements. At lower power levels, transistors or other semiconductor switches could be used.

Initially Th1 is triggered on. The voltage across this device falls to almost zero. A current accompanied by an appropriate magnetic flux will rise in the top half of the transformer primary winding. Since this flux is common to both halves of the transformer winding, the source voltage E_s would be induced in the lower half and the commutating capacitor would have a voltage of $2E_s$ across it with the polarity shown. Subsequently, when Th2 is triggered on, the commutating capacitor applies a voltage of approximately $-2E_s$ to appear across Th1. When this reverse voltage is applied for a sufficient length of time across Th1, it will be turned off. Th2 will now be conducting and a voltage of $2E_s$ will appear across the transformer primary and the commutating capacitor, but with a reverse polarity to that shown in Figure 8.1. At the next trigger pulse Th1 will be again turned on and Th2 off. Thus, if trigger pulses are periodically applied to the alternate thyristors, an approximately rectangular voltage wave will be obtained at the transformer output terminals.

8.2 Analysis of the Circuit Shown in Figure 8.1

Assuming that Th1 is conducting, the switch has the position shown, and the instantaneous conditions may be expressed by

$$E = E_s - E_{Th1} = L\frac{di_1}{dt} + \frac{V_c}{2} \tag{8.1}$$

where E_{Th1} = drop across Th1

If a purely resistive load is assumed, then $V_2 = i_2R_L$ and

$$V_c = i_2R_L\frac{N1}{N2} \tag{8.2}$$

where N2 is the secondary turns and N1 is the primary turns.

Also,

$$i_2N2 = (i_1 - i_c)N1/2 - i_cN1/2$$

$$= (i_1 - 2i_c)N1/2 \tag{8.3}$$

and

$$i_c = C\frac{dV_c}{dt} \tag{8.4}$$

Substituting for i_c in Eq. 8.3,

$$i_2N2 = \left(i_1 - 2C\frac{dV_c}{dt}\right)N1/2$$

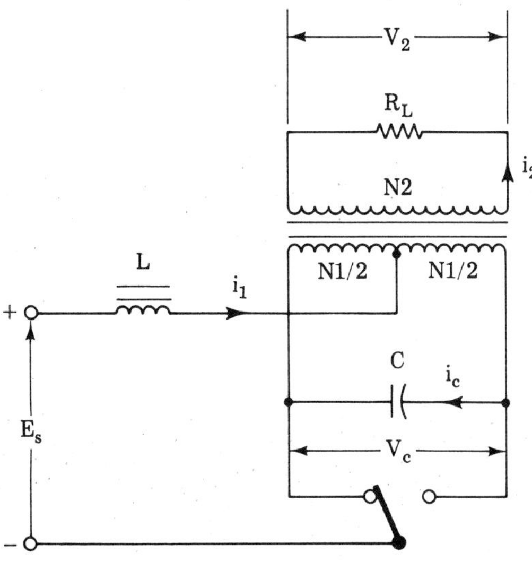

Figure 8.2. Analysis for circuit of Figure 8.1.

and $$i_2 = \left(i_1 - 2C\frac{dV_c}{dt}\right)\frac{N1}{2N2}$$

In Eq. 8.2, substituting the above value of i_2,

$$V_c = \left(i_1 - 2C\frac{dV_c}{dt}\right)\frac{R_L}{2}\left(\frac{N1}{N2}\right)^2 \tag{8.5}$$

If the value of L is such that i_1 is substantially constant at a value I, then Eq. 8.5 gives a *first-order equation*; i.e.,

$$\frac{dV_c}{dt} + \frac{n^2}{CR_L}V_c = \frac{I}{2C}$$

and $$V_c = \frac{IR_L}{2n^2} + Ae^{-n^2t/CR_L}$$

where $n = N2/N1$ and A is a constant that may be determined as follows: V_c at the beginning of a second half period has the same magnitude as at the beginning of a period, but with reverse sign. Hence,

$$\frac{IR_L}{2n^2} + A\epsilon^{-n^2T/2/CR_L} = -\frac{IR_L}{2n^2} - A$$

where T is the perodic time. From which

$$A = \frac{-IR_L/n^2}{1 + \epsilon^{-n^2T/2/CR_L}}$$

Thus,

$$V_c = \frac{IR_L}{2n^2}\left(1 - \frac{2e^{-n^2t/CR_L}}{1 + e^{-n^2T/2/CR_L}}\right)$$

In the above equation $n = N2/N1$, but if N1 is replaced by one-half the total primary turns N1′, then $IR/2n^2$ becomes $2IR/(N2/N1')^2$. This is the load resistance reflected across one-half of the primary winding and I is the current in this load when $C[(dV_c)/dt] = 0$.

Thus, $$2IR_L/(N2/N1')^2 = 2E$$

and $$V_c = 2E\left(1 - \frac{2e^{-n^2t/CR_L}}{1 + e^{-n^2T/2/CR_L}}\right) \tag{8.6}$$

Equation 8.6 indicates that V_c increases exponentially towards a limiting voltage 2E. Practical values of $(n^2T/2)/CR_L$ are generally such that

$$e^{-n^2T/2/CR_L} \ll 1$$

and hence $$V_c = 2E(1 - 2e^{-n^2t/CR_L}) \tag{8.7}$$

8.3 Design of Commutating Capacitor

The voltage across the commutating capacitor is given by

$$V_c = 2E(1 - 2e^{-n^2t/CR_L}) \tag{8.8}$$

when $V_c = 0$, let $t = t_o$, where t_o = turn-off time of the SCR. From Eq. 8.8,

$$0 = 2E - 4Ee^{-n^2t_o/CR_L}$$

or

$$e^{-n^2t_o/CR_L} = \frac{1}{2}$$

$$\frac{n^2t_o}{CR_L} = \log e^2 = .6931$$

Therefore,

$$t_o = \frac{.6931 \times CR_L}{n^2}$$

The value of C, R_L, and n^2 must be chosen such that t_o is greater than the turn-off time of the SCRs. The example below illustrates the design of the parallel capacitor commutated inverter and also the various waveforms in the circuit.

EXAMPLE 8.1

In the inverter circuit shown in Figure 8.1, the battery voltage E_s = 240V, $\frac{N1}{2}:N2 = 1:1$. For a load resistance, R_L = 5 ohms, write the expression for voltage on the capacitor and find the necessary value of C to obtain 40 μsec turn-off time on the thyristor. Assume inductor L is very large and the transformer is ideal. Sketch the following waveforms in the above circuit with time, t:

a. capacitor voltage V_c
b. voltage across Th1
c. capacitor current, i_c
d. current in Th2, i_2

Show also on the sketch triggering points on Th1 and Th2.

SOLUTION

The voltage across the capacitor is given by the expression

$$V_c = 2E(1 - 2e^{-n^2t/CR_L})$$

where $N2/N1/2 = 1$

$$\therefore n = \frac{N2}{N1} = \frac{1}{2}$$

so $$V_c = 2E(1 - 2e^{-t/4CR_L})$$

or $$t_o = \frac{.6931CR_L}{n^2}, \text{ when } V_c = 0$$

Therefore, $$C = \frac{40 \times 10^{-6}}{.6931 \times 4 \times 5} = 2.88\mu F$$

The various waveforms are sketched in Figure 8.3.

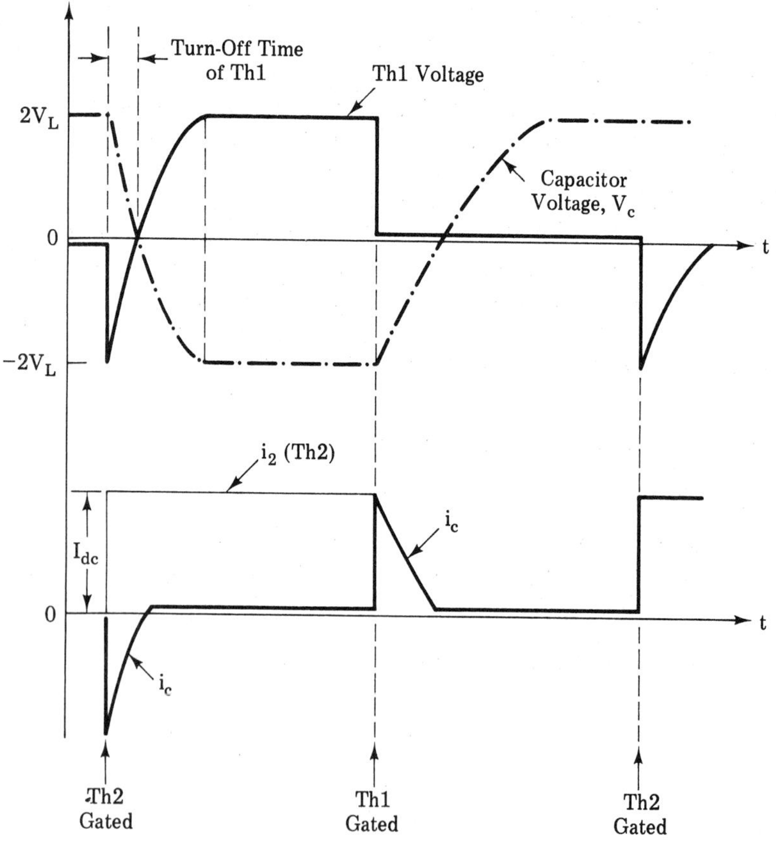

Figure 8.3. Waveforms of capacitor and thyristor voltage and currents over one cycle of operation.

8.4 Three-Phase Inverter

The three-phase inverter circuit changes dc input voltage to a three-phase variable-frequency variable voltage output. The input dc voltage may be from a dc source or a rectified ac voltage. Most inverter applications require a means of voltage and frequency control.

The methods of control for voltage can be grouped into three broad categories:

1. Control of voltage supplied to the inverter.
2. Control of voltage within the inverter.
3. Control of voltage delivered by the inverter.

8.5 Control of Voltage Supplied to the Inverter

The control of voltage supplied to the inverter may be varied in several ways. If an ac voltage is available the most commonly used circuit is the phase-controlled rectifier. If the available voltage is dc, then a chopper circuit is used to vary the voltage to the input of the inverter.

The principal advantages of voltage-control schemes, where the dc supply voltage to the inverter is controlled, are given below:

1. The inverter output voltage waveshape and its harmonic content are not significantly changed as the voltage is controlled.
2. Voltage delivered to the inverter is easily controlled.

The main disadvantages of the above scheme are as follows:

1. The commutating voltage in many inverters is proportional to the dc input voltage. This makes the inverter current capability decrease as the dc voltage is reduced. Therefore, the controlling of the dc voltage to such inverters is not desirable when a large variation in the output voltage is required and when high load current is necessary at the reduced voltages. If these inverters are designed to realiably commutate the highest currents at a reduced dc voltage then, at the high dc input voltages, there is excessive commutating voltage which, generally, produces increased circulating currents resulting in higher losses. This problem could be overcome by providing a fixed dc bus of constant commutating capacity.

2. The power delivered by the inverter is handled twice, once by the dc voltage control and once by the inverter.
3. Most efficient dc voltage control schemes require filtering in the dc circuit. This causes slower response time in a complete closed-loop voltage regulated inverter power supply.

8.6 Control of Voltage Within the Inverter

An excellent method of controlling the voltage within an inverter involves the use of pulse modulation techniques. In essence, with this technique the inverter output voltage is a pulse width modulated wave, and the voltage is controlled by varying the duration of the output voltage pulses.

8.7 Control of Voltage Delivered by the Inverter

The output voltage of the inverter may be controlled by using a transformer at the output. For wide frequency range this method is not a practical solution.

8.8 The Basic Three-Phase Inverter Circuit

The inverter power circuits used in ac variable-speed drive systems have been classified into two groups:

1. Adjustable voltage input (AVI).
2. Pulse width modulation (PWM).

Adjustable Voltage Input Inverter

The AVI inverter consists of three major circuits:

a. a phase-controlled rectifier for voltage control
b. an inverter for frequency control
c. a fixed dc bus to provide constant commutating capability.

A block diagram of this circuit is shown in Figure 8.4. The phase-controlled bridge has been described in detail in Chapter 4. However,

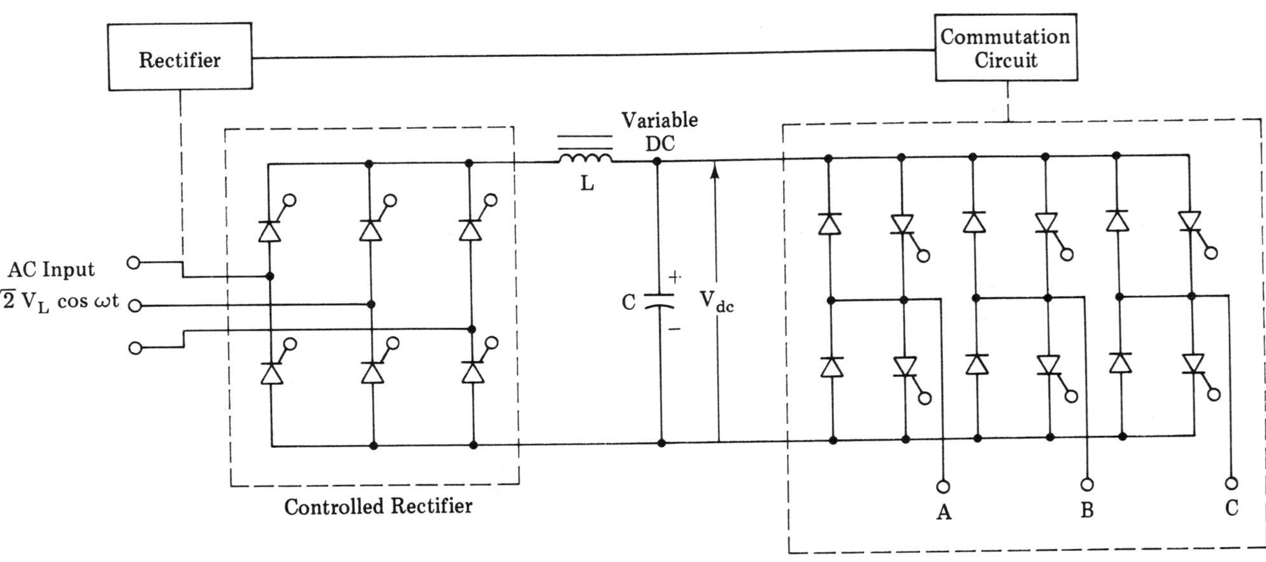

Figure 8.4. AVI with phase-controlled rectifier.

one characteristic of the phase-controlled bridge is important in operation of the AVI inverter. The input displacement factor of the phase-controlled bridge varies with the output controlled voltage. At low output voltage of the converter, the displacement factor is poor, which means power factor at the input is low.

An output filter is required to smooth the ripple in the output voltage of the controlled rectifier. The ripple voltage at the output of the controlled rectifier is worst at low dc bus voltage at a constant output current. Discontinuous current conduction is possible at low output voltage. This condition is to be prevented by large inductance at the output of the controlled rectifier.

The inverter bridge can deliver power or regenerate and consists of six thyristors which alternately connect each phase to positive and negative dc bus. Each thyristor is on for 180° and the switching sequence produces a three-phase output voltage. The waveform (Figure 8.5) at the output is referred to as a six-step waveform. The waveshape remains constant at all frequencies. The output frequency is controlled by the inverter and the amplitude V_{DC} is controlled by the phase-controlled bridge. These two variables must be in a proper ratio to keep the ratio of voltage to frequency constant.

A three-phase full-wave diode bridge and a chopper could also be used to supply the variable dc voltage to the input of the inverter as shown in Figure 8.6. This circuit has a good power factor at the ac input. If an ac source is not available, the chopper is also capable of operating from a dc source.

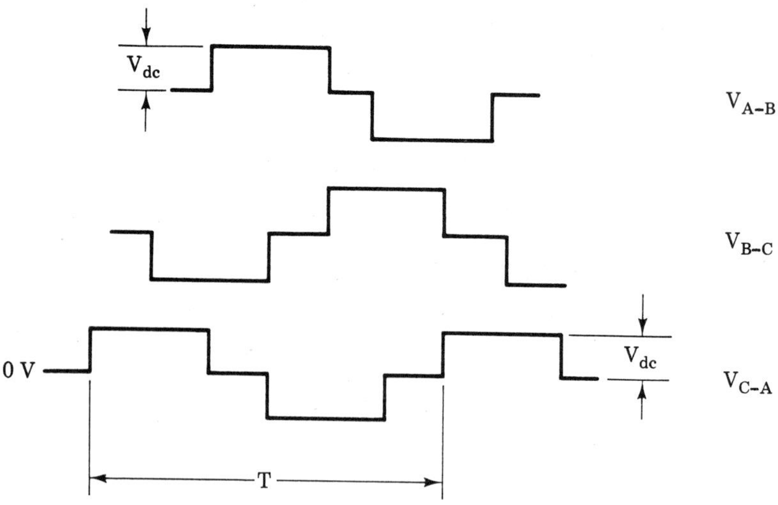

Figure 8.5. The waveform at the inverter output.

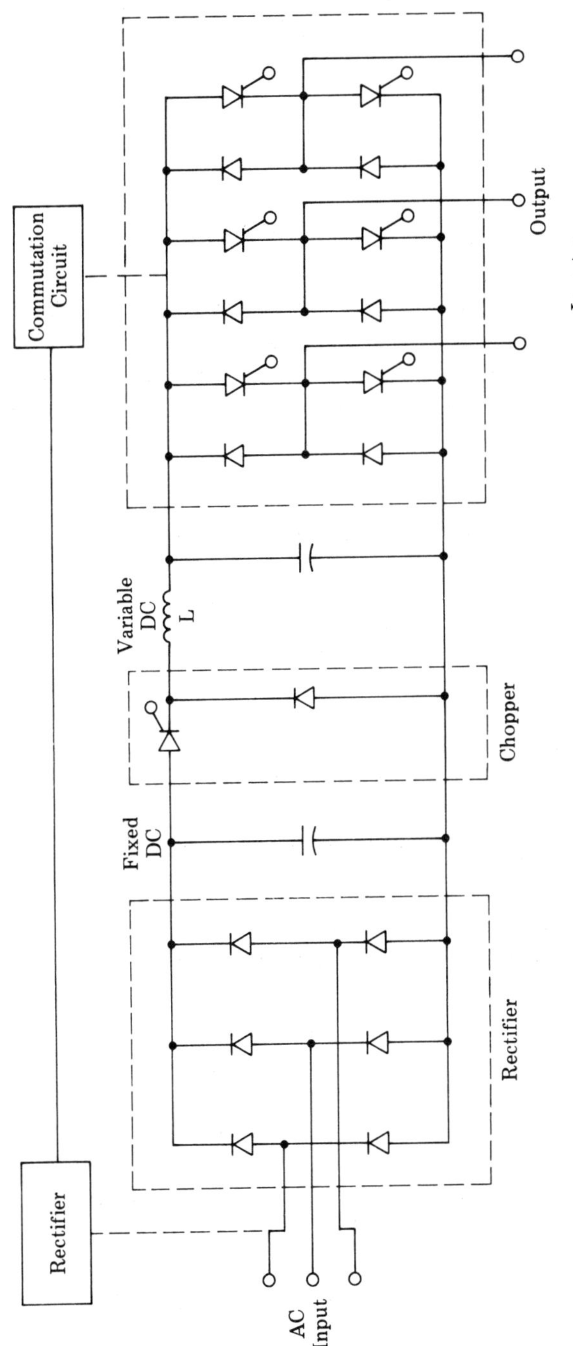

Figure 8.6. AVI with chopper input.

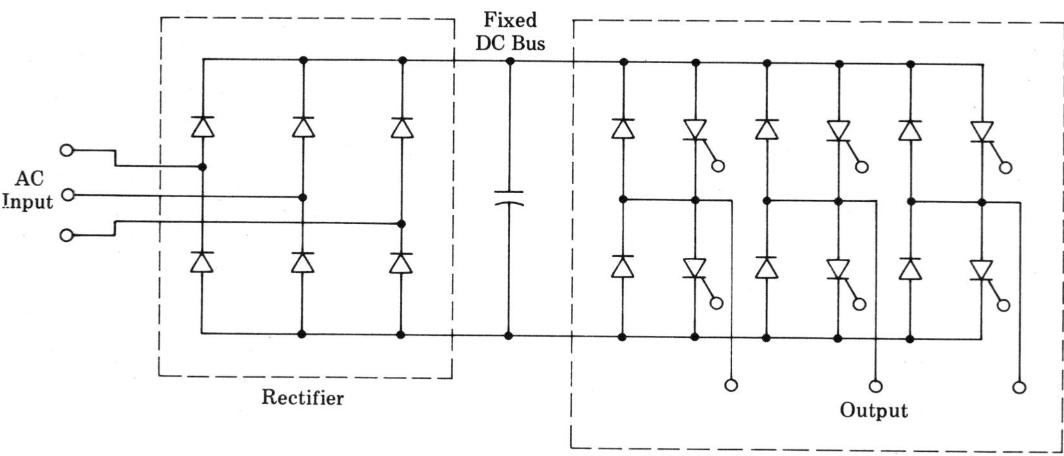

Figure 8.7. Pulse width modulated inverter.

Pulse Width Modulated Inverter (PWM)

In the PWM inverter (Figure 8.7) the voltage and frequency control are accomplished with one power circuit and proper control logic. The output voltage waveform is of constant amplitude whose polarity reverses periodically to provide the output fundamental frequency. The output voltage is varied through pulse width control. The filtering of the output voltage is partially accomplished by the motor inductance. The chopping frequency of the output voltage is usually referred to as the carrier. It is desirable to keep the ratio of carrier to motor frequency as high as possible so that any additional motor losses due to harmonics of the carrier will be minimized. But too high a carrier frequency increases the commutation and other losses in the inverter. Hence a carrier frequency has to be chosen so that motor losses are kept to a minimum and the losses within the inverter are not too high. The control logic circuit of a PWM inverter is much more complicated than an inverter which is accomplishing only frequency control. The complicated control logic circuit is a distinct disadvantage of the PWM inverter. The PWM inverter could be designed to obtain square-wave or sine-wave modulation as shown in Figure 8.8.

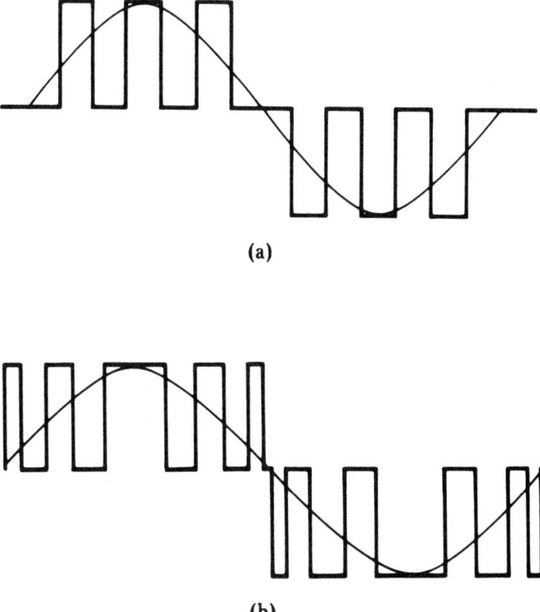

Figure 8.8. PWM inverter modulation: (a) square wave modulation; (b) sine wave modulation.

8.9 Analysis of the Three-Phase Full-Wave Bridge Circuit of Figure 8.4

As discussed in Chapter 4, the output voltage of the controlled rectifier is given by

$$V_{dc} = \frac{1}{2\pi/6} \int_{-\pi/6+\alpha}^{\pi/6+\alpha} \sqrt{2}\, V_L \cos \omega t \; d(\omega t) \tag{8.9}$$

assuming continuous conduction, α = delay angle and V_L = line-to-line input voltage.

Therefore,

$$V_{dc} = \frac{3\sqrt{2}\, V_L}{\pi} \cos \alpha \tag{8.10}$$

As seen from Eq. 8.10, the output dc voltage varies as a function of delay angle α.

8.10 Line-to-Line Voltage Output of the Inverter

The line-to-line and line-to-neutral voltage outputs of the inverter are shown in Figure 8.9.

Let V_{rms} = rms value of the line-to-line voltage output of the inverter.

Then,

$$V_{rms} = \sqrt{\frac{1}{\pi} \int_{0}^{2\pi/3} V_{dc}^{2} \, d\theta} = \sqrt{\frac{2}{3}}\, V_{dc} \tag{8.11}$$

Also from Figure 8.9, the line-to-line voltage of the inverter may be written as

$$V_{A\text{-}B} = \sqrt{2}\, V_1 \cos \omega t + \sqrt{2}\, V_5 \cos 5\omega t + \sqrt{2}\, V_7 \cos 7\omega t \tag{8.12}$$

In the above equation the even harmonics are excluded because the waveform is symmetrical about $\omega t = 0$. Triple harmonics are zero in a balanced three-phase system. $V_1, V_5, V_7, \ldots$ in Equation 8.12 are the rms values of the fundamental, fifth harmonic, seventh harmonic, etc. of output.

Fundamental component V_1 is given by

$$V_1 = \frac{1}{\sqrt{2}\,\pi} \int_{0}^{2\pi} V_{A\text{-}B} \cos \omega t \; d(\omega t)$$

$$= \frac{1}{\sqrt{2}} \frac{4}{\pi} \int_{0}^{\pi/3} V_{dc} \cos \omega t \; d(\omega t) = \frac{\sqrt{6}}{\pi} V_{dc} \tag{8.13}$$

The rms value of the fundamental is important in ac motor control

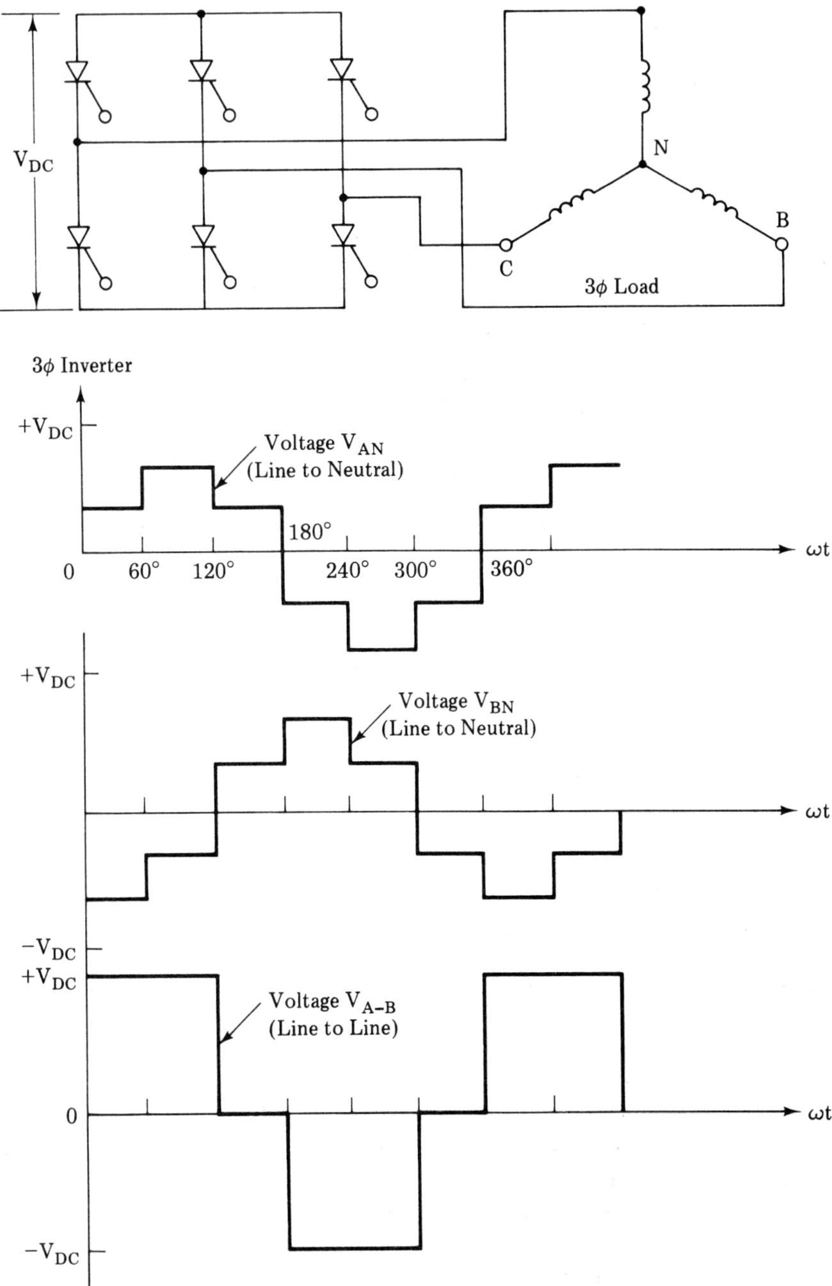

Figure 8.9. Three-phase inverter waveform for 180° conduction.

applications. The rms value of the fifth harmonic component is

$$V_5 = \frac{4}{\sqrt{2}\pi}\int_0^{\pi/3} V_{dc} \cos 5\omega t \, d(\omega t)$$

$$= \frac{-1}{5}\frac{\sqrt{6}}{\pi} V_{dc} \tag{8.14}$$

It is seen from Eq. 8.14 that the magnitude of the 5th harmonic is 20 percent of the fundamental and is negative.

The rms value of the 7th harmonic component is

$$V_7 = \frac{4}{\sqrt{2}}\int_0^{\pi/3} V_{dc} \cos 7\omega t \, d(\omega t)$$

$$= \frac{1}{7}\frac{\sqrt{6}}{\pi} V_{dc} \tag{8.15}$$

It is seen from Eq. 8.15 that the magnitude of the 7th harmonic is 1/7 of the fundamental and is positive. The magnitudes of the harmonic components are very important in the control of induction motors as is explained in Chapter 9.

8.11 Commutation Circuits for Inverters and Choppers

In dc link inverters and chopper circuits, the thyristors will not turn off naturally as they are operating from dc input voltage. The thyristors must be turned off by what are known as commutation circuits. Classification of commutation methods is given in the next section. There are few, if any, short cuts in commutation; commutation should be a major consideration in the economic design of inverter circuits. Other than by reducing turn-off time in thyristors, the problem will not change significantly.

8.12 Classification of Commutation Methods

There are six basic methods of commutation by which thyristors may be turned off as given below:

Class A—Commutation by Load Resonance

This type of commutation circuit using L-C components in series with the load is shown in Figures 8.10(a) and 8.10(b).

In Figure 8.10(a), load R_L is in parallel with the capacitor and in Figure 8.10(b), load R_L is in series with L-C circuit. The underdamped

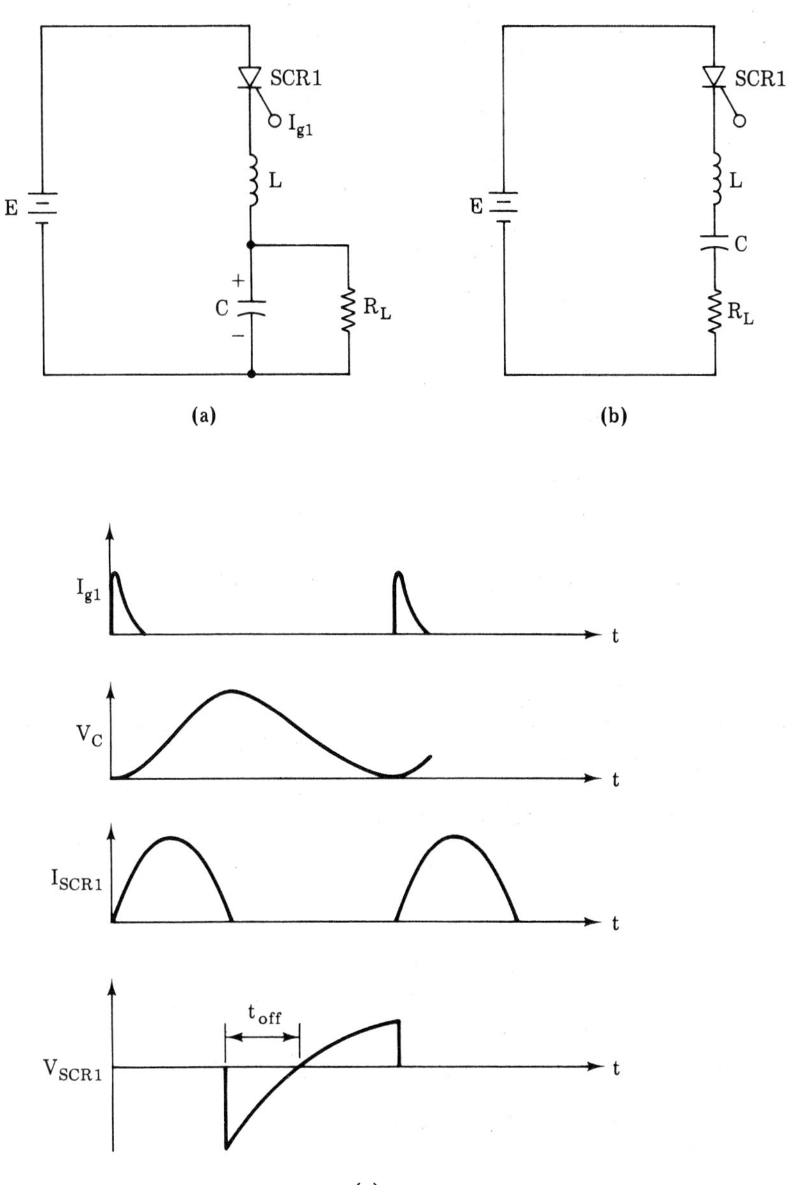

Figure 8.10. Class A commutation circuit: (a) load in parallel with capacitor; (b) load in series with capacitor; (c) waveforms.

L-C resonating circuit in series with the load applies a reverse voltage to SCR1 to turn it off as shown in Figure 8.10(c). The capacitor supplies the commutation energy. The period of L-C resonating circuit and the switching rate of SCR1 determines the turn-off time. This type of commutation is used in series inverter circuits. The commutation process for Figure 8.10(a) is shown in Figure 8.10(c). When SCR1 is triggered by injecting gate current I_{g1}, the capacitor charges up to a voltage

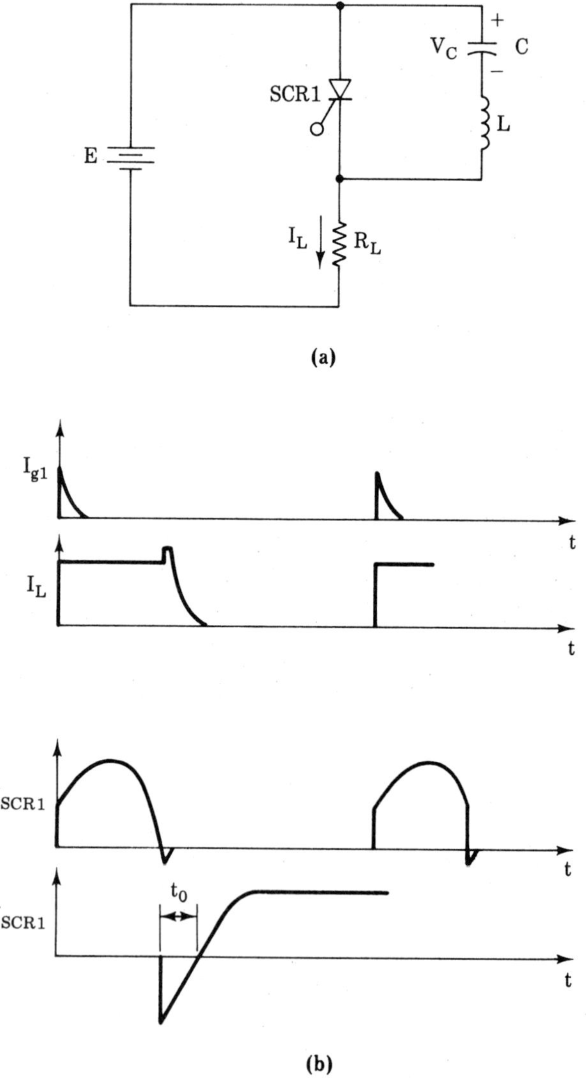

Figure 8.11. Class B commutation method: (a) circuit; (b) waveforms.

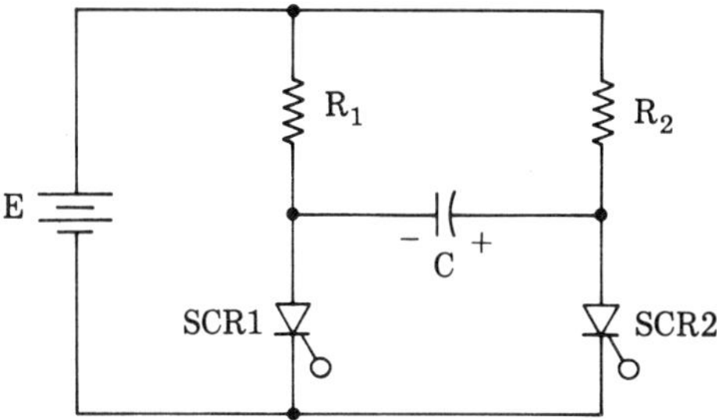

Figure 8.12. Class C commutation circuit.

higher than E, depending on losses in the circuit. The current I_{SCR1} falls to zero and SCR1 is reverse biased and therefore the SCR is turned off.

Class B—Self-Commutated by an LC Circuit

In this method, the LC resonating circuit is across the SCR and not in series with the load. The commutating circuit is shown in Figure 8.11(a), and the associated waveforms are shown in Figure 8.11(b).

The capacitor C is initially charged to a voltage V_c with the polarity shown. When SCR1 is triggered, the capacitor discharges through the LC resonant circuit and a reverse voltage is applied across the SCR, turning it off. Once the SCR is turned on, it conducts for a definite period, and then it automatically turns off.

In the Class B commutation method, the commutating component does not carry the load current. This method of commutation is used in dc chopper circuit.

Both Class A and B turn-off circuits are self-commutating types, i.e., in both of these circuits, the SCR turns off automatically after it has been turned on.

Class C—Switching a Charged Capacitor by a Load-Carrying SCR

The Class C commutation circuit is shown in Figure 8.12. In this circuit, the conducting SCR is turned off when it is reverse biased by switching the capacitor voltage across it through the second SCR.

In Figure 8.12, let us assume that SCR1 is conducting, so current is flowing through R1. At this instant SCR2 is blocking. So capacitor C is charged to the source voltage E through R2. When SCR2 is turned on, the capacitor is switched across SCR1, which applies a reverse voltage across SCR1 and turns it off.

Figure 8.13 shows the model of the circuit when SCR1 is conducting and SCR2 is forward blocking. The capacitor is charged to

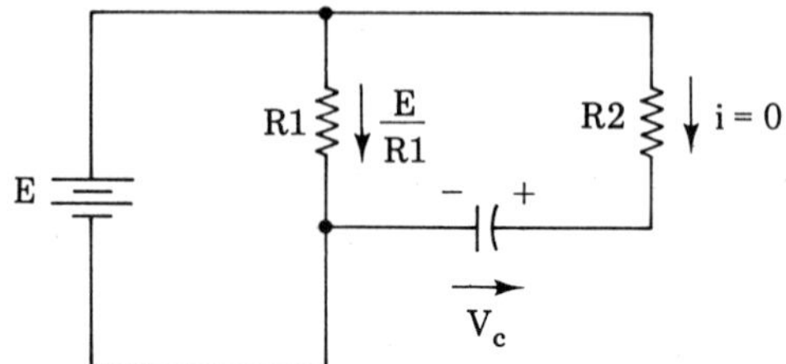

Figure 8.13. Circuit of Figure 8.12 when SCR1 is conducting.

voltage E with the polarity shown. When SCR2 is turned on, V_c is applied across SCR1 (see Figure 8.14), turning off SCR1.

Capacitor C discharges through SCR1 and R1. When SCR1 turns off, the capacitor charges through R1 with a time constant R1C to the supply voltage E, but with opposite polarity to that shown in Figure 8.13. SCR2 is now conducting through R2 and SCR1 is forward blocking. The circuit is now ready to commutate SCR2 by switching on SCR1. The design criteria for the circuit are as follows:

1. The maximum dV/dt across SCR1 gives one design equation, i.e.,

$$\frac{2E}{R1C} < \frac{dV}{dt}\,(\max) \tag{8.16}$$

   The maximum allowable dV/dt for SCR1 may be obtained from the SCR data sheet.

2. The capacitor must be sufficiently large to satisfy the turn-off time, t_o, of SCR1 as shown below:

$$V_c = E(1 - 2e^{-t/R1C}) \tag{8.17}$$

   when $V_c = 0$, let $t = t_o$, where t_o = turn-off time of the SCR or $0 = 1 - 2\,e^{-t/R1C}$

$$\therefore t_o = .6931\ R1C \tag{8.18}$$

So, from Eq. 8.18, R1 and C must be such that t_o, the turn-off time of the SCR, is satisfied. The waveforms at various points on the commutation circuit are shown in Figure 8.14(b).

EXAMPLE 8.2

In the circuit of Figure 8.12 the source voltage E = 120V, and the current through loads R1 and R2 is 20A. The turn-off time of both the SCRs is 60 μsec. Find the value of C for successful commutation.

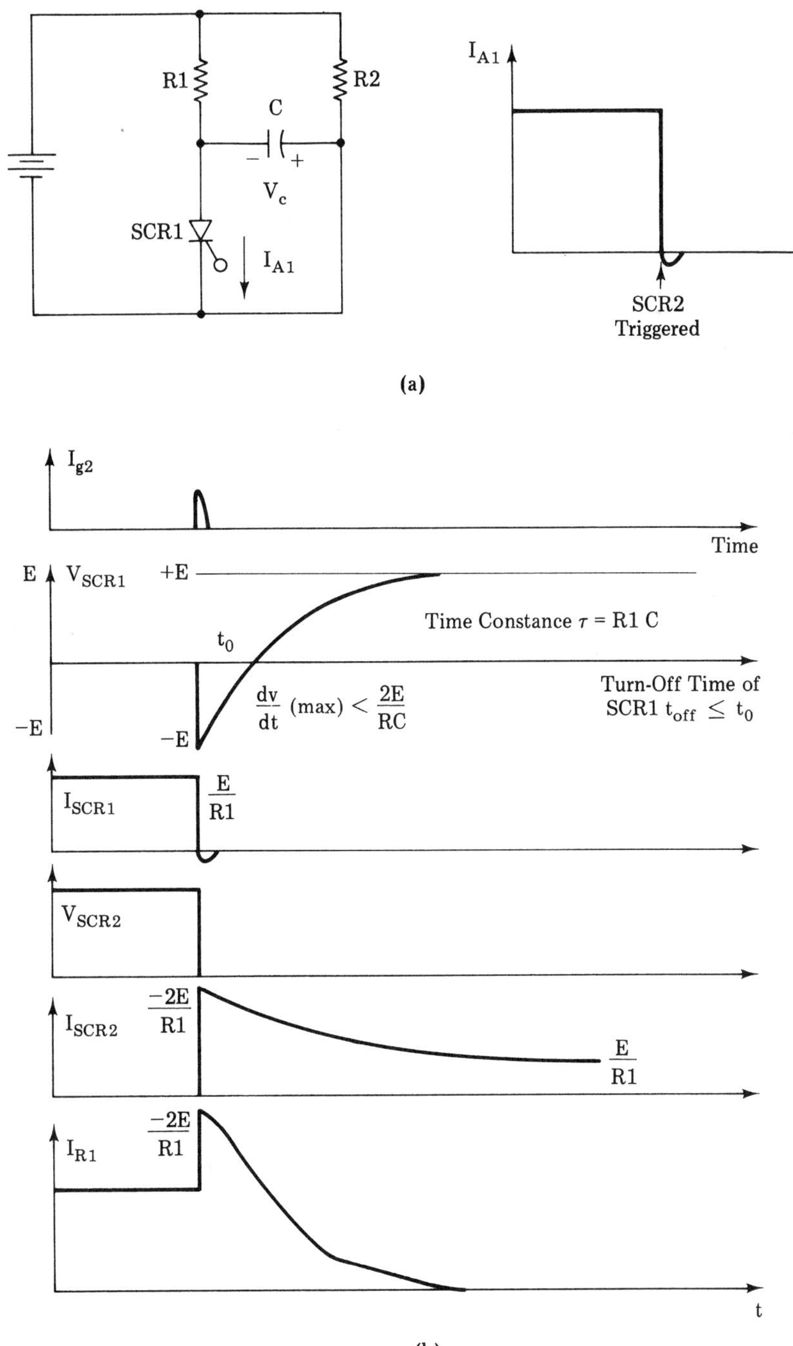

Figure 8.14. Model of Figure 8.12 when SCR2 is triggered: (a) circuitry; (b) waveforms at various points in the circuit.

SOLUTION

The resistances $R1 = R2 = \frac{120}{20} = 6\Omega$. From Eq. 8.18 t_o = turn-off time of SCR = .6931 R1C.

$$\therefore C = \frac{60 \times 10^{-6}}{.6931 \times 6} = 14.4\mu F$$

Class D—An Auxiliary SCR Switching a Charged Capacitor

A typical class D commutation circuit is shown in Figure 8.15(a), and the corresponding waveforms are shown in Figure 8.15(b). In this circuit, SCR1 is the main load-carrying component and SCR2 is the auxiliary device.

Initially SCR2 is triggered, so that capacitor C is charged to a voltage v_c with the polarity shown. SCR2 turns off when the capacitor is fully charged and the current through SCR2 falls below the holding current.

When SCR1 is triggered, the load current I_L flows through SCR1 and R_L and the commutating current flows through SCR1, L, D, and C, and charges up the capacitor to the opposite polarity and is ready for commutating SCR1. When SCR2 is triggered, the capacitor voltage v_c reverse biases SCR1 and turns it off.

Design of the Commutation Circuit

The basic design of the commutation circuit involves the proper choice of capacitor C and inductor L.

Commutating Capacitor. The magnitude of the commutating capacitor is dependent on the following circuit parameters:

1. Maximum load current to be commutated, I_L.
2. Turn-off time, t_o, of SCR1.
3. The battery voltage, E.

The turn-off time, t_o, of SCR1 is known from the manufacturer's data sheet. The capacitor voltage changes from $-E$ to 0 during turn-off time, t_o. Assuming load current, I_L, remains constant during turn-off time, t_o,

$$CE = I_L t_o \tag{8.19}$$

$$\therefore C = \frac{I_L t_o}{E} \tag{8.20}$$

Commutating Inductor L. The design of the inductor L is dependent on two contradictory criteria as follows:

1. The acceptable maximum capacitor current I_c when SCR1 is fired.
2. The time interval $(t_2\text{-}t_1)$, during which capacitor voltage must reset to correct polarity for commutating SCR1.

The capacitor current I_c is the oscillatory current through SCR1, L, D, and C when SCR1 is triggered. The peak value of current I_c is given by the expression

$$I_c\big|_{peak} = \frac{E}{\omega_r L} \tag{8.21}$$

where ω_r = oscillating frequency $= \dfrac{1}{\sqrt{LC}}$ in rad/sec. From Eq. 8.21,

$$I_c\big|_{peak} = E\sqrt{\frac{C}{L}} \tag{8.22}$$

Also,

$$T_r = \frac{2\pi}{\omega_r} = 2(t_1 - t_2) \tag{8.23}$$

where T_r = periodic time during oscillation.

Let the maximum current through SCR1 be limited to I_{Lm}. From Eq. 8.22,

$$E\sqrt{\frac{C}{L}} \leq I_{Lm}$$

or

$$L \geq C\left(\frac{E}{I_{Lm}}\right)^2 \tag{8.24}$$

The resetting time could be reduced by decreasing the value of L, but the peak capacitor current would increase as seen from Eq. 8.22. A large resetting time would limit the minimum voltage available at the load, which means the range of voltage available at the load is reduced.

The minimum load voltage available is given by

$$V_o\big|_{min} = \frac{t_1 - t_2}{T} E \tag{8.25}$$

where T is the chopper time period, or,

$$V_o\big|_{min} = \frac{\pi\sqrt{LC}}{T} \cdot E \tag{8.26}$$

Therefore,

$$L \leq \left(\frac{V_o|_{min}}{E}\right)^2 \frac{T^2}{\pi^2 C} \tag{8.27}$$

EXAMPLE 8.3

In Figure 8.15(a), the battery voltage E = 50V, the maximum load current, I_L = 50A, t_o, turn-off time of SCR1 = 30 μsec, chopper frequency is 500 Hz. Calculate the values of the commutating capacitor, C, and the commutating inductor, L. The load voltage variation required is 10% to 100%.

SOLUTION

For reliable commutation, assume 50% tolerance on turn-off time. Hence t_o = 45 μsec. From Eq. 8.20,

$$C = \frac{50 \times 45 \times 10^{-6}}{50} = 45\mu F$$

Period T = 2 msec.
From Eq. 8.27,

$$L \leq (\cdot 1)^2 \frac{(2 \times 10^{-3})^2}{\pi^2 \times 45 \times 10^{-6}}$$

$$\leq 90\mu H$$

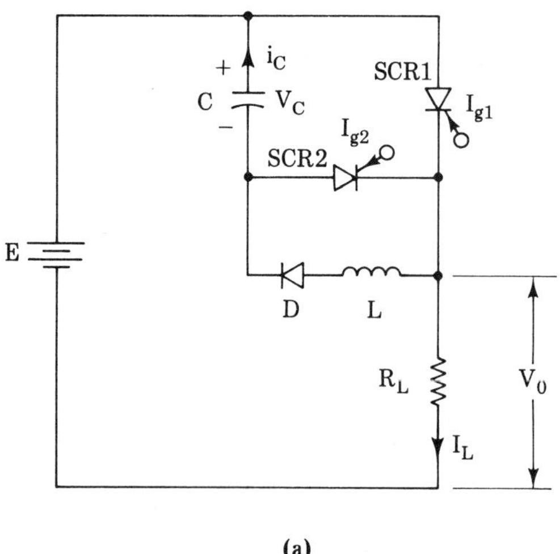

(a)

Figure 8.15. Class D commutation: (a) circuit. (*Continued*)

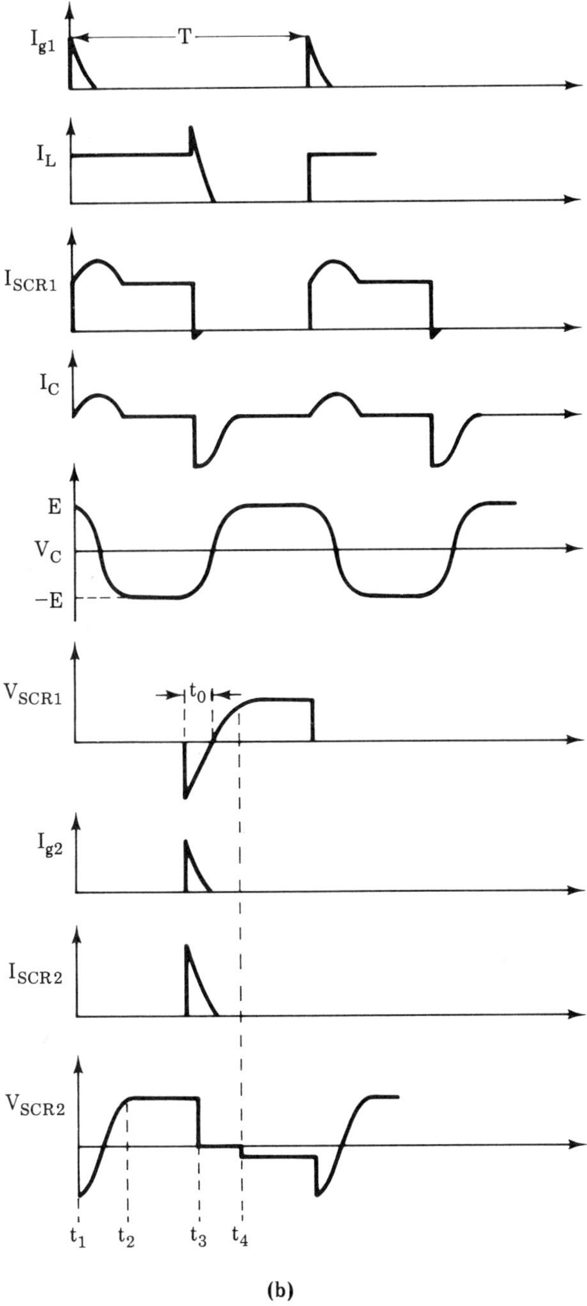

(b)

Figure 8.15. (b) waveforms.

From Eq. 8.24

$$L \geq 45 \times 10^{-6} \left(\frac{50}{50}\right)^2$$

$$\geq 45\mu H$$

The range of commutating inductor is

$$45\mu H < L < 90\mu H$$

The choice of lower value of L would allow a larger voltage variation at the load.

Class E—An External Pulse Source for Commutation Energy

This class of commutation circuit uses an external pulse source to reverse bias the current-carrying SCR and thereby turn it off. The width of the pulse has to be chosen such that the SCR is reverse biased for a period greater than the turn-off time of the SCR. A typical Class E commutation circuit is shown in Figure 8.16(a), and the associated waveforms are shown in Figure 8.16(b). In this particular case, the commutating pulse is applied in series with the SCR. The transformer must be capable of carrying the load current.

When SCR1 is triggered into conduction by injecting current I_{g1} to its gate, the load current I_L flows through SCR1, secondary of the pulse transformer, and the load R_L. To turn off SCR1, a positive pulse

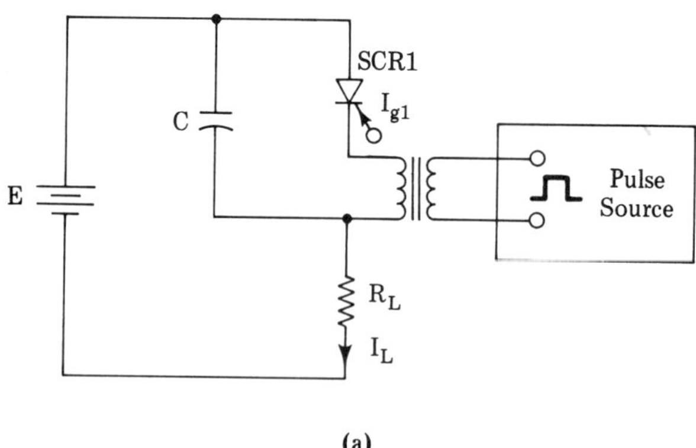

(a)

Figure 8.16. Class E commutation: (a) circuit. (*Continued*)

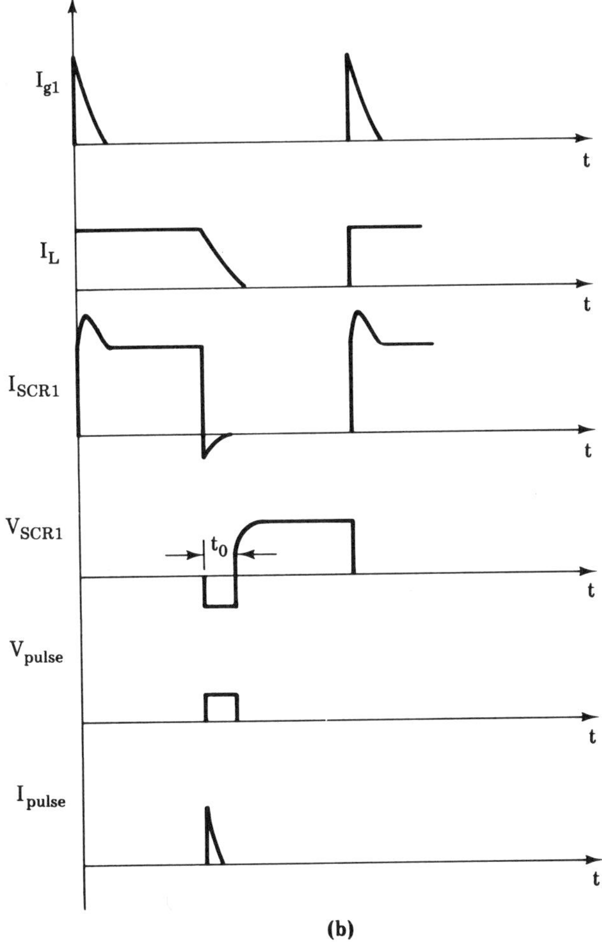

(b)

Figure 8.16. (b) waveforms.

from the pulse transformer is applied to the cathode of SCR1, which turns it off.

Class F—AC Line Commutation

The load-carrying SCR1 is reverse biased when the ac source voltage reverses polarity. A typical line commutated circuit and its associated waveforms are shown in Figures 8.17(a) and 8.17(b).

The SCR is reverse biased for a period of t_o as shown in the figure. The maximum frequency at which this circuit can operate depends

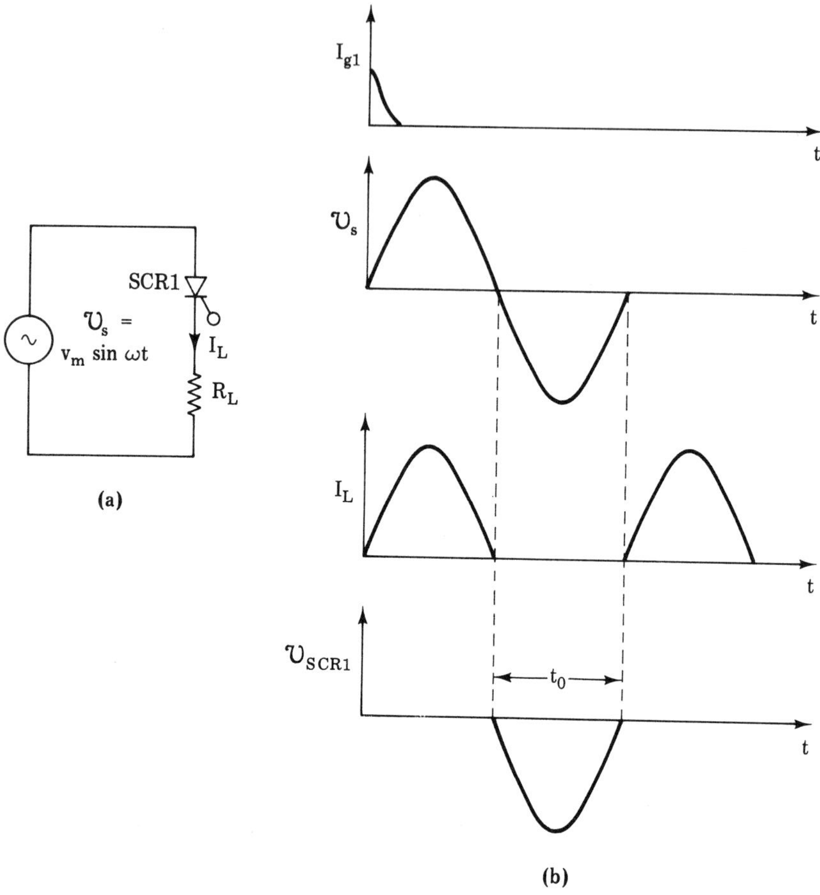

Figure 8.17. Class F commutation: (a) circuit; (b) waveforms.

on the turn-off time of the SCR. It is obvious from Figure 8.17(b) that t_o must be greater than the turn-off time of the SCR selected.

8.13 Impulse-Commutated Inverter

In the pulse width modulated inverter (PWM) the thyristor switches are operated at high frequency, to eliminate the lower order of harmonics present in the output waveform. Such types of inverters require lossless commutation circuits and short clearing time between two consecutive commutations. Many circuits have been published which could be used as commutation circuits in inverters. The modified McMurray inverter and the McMurray-Bedford circuit are discussed in the following sections.

8.14 Modified McMurray Inverter

The modified version of the McMurray inverter is shown in Figure 8.18. This is the auxiliary impulse-commutated (McMurray) circuit with the addition of two diodes (D_{1A} and D_{2A}), and a resistor (R) for return of capacitor overcharge to the supply. The operation and design of the power circuit is given in this section.

Assume main SCR1 is conducting with some load current I_L. The commutating capacitor (C_c) is charged with the polarity as shown to a voltage V_c. To commutate main SCR1, the auxiliary SCR1A is turned on. This completes the LC circuit and an oscillation is initiated. As soon as I_c (commutating current) exceeds I_L, the excess current is flowing through the feedback diode D_1, thus providing a reverse bias on SCR1 sufficient to turn it off. At the end of the commutation process, the capacitor voltage V_c will be charged in the opposite direction. At this point SCR2 will be carrying load current, and it can be turned off by triggering SCR2A to initiate the commutation process.

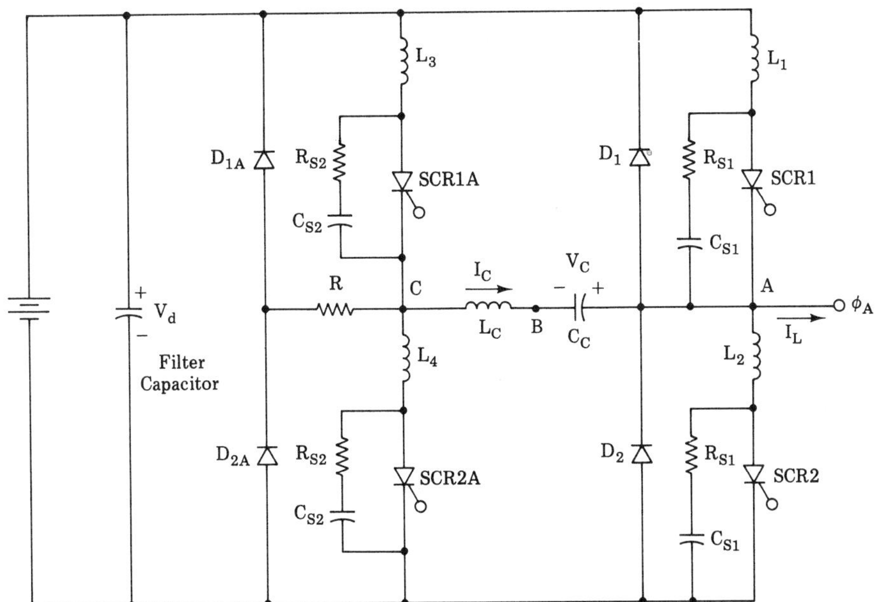

Figure 8.18. One phase of modified McMurray inverter circuit.

8.15 Operation of the Commutation Circuit With Lagging Load

The circuit waveforms and design are explained assuming a lagging load in Figure 8.19.

Interval A. SCR1 is conducting and the load current I_L is flowing through SCR1.

Interval B. Auxiliary SCR1A is triggered to turn off main SCR1. Discharge current I_c of commutating capacitor forces current through SCR1, turning it off. The oscillatory circuit is SCR1A, L_c, C_c, SCR1, L_1, and L_3.

Interval C. The load current remains constant due to inductance of the load. As the discharge current I_c exceeds the load current, excess current flows through feedback diode D_1, thus reverse biasing SCR1 and turning it off. The discharge current reaches a maximum value

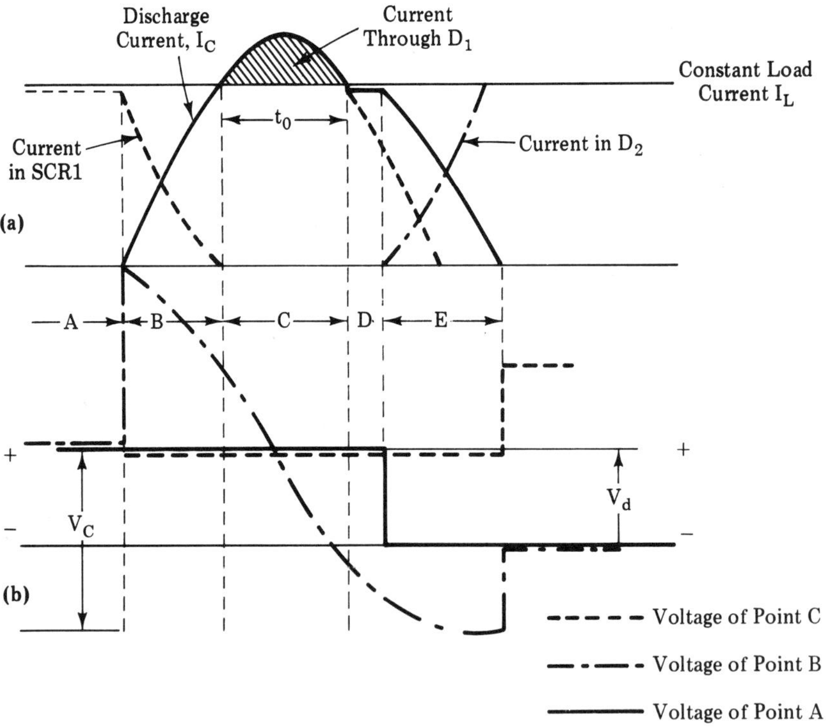

Figure 8.19. Operation of the commutation circuit with lagging load: (a) current waveform; (b) voltage waveform.

when the capacitor voltage is zero, and then decreases as the capacitor charges in the reverse direction.

Interval D. As discharge current, I_c, falls below load current, I_L, the diode D_1 ceases to conduct. The load continues to draw a constant current through auxiliary SCR1A, L_c, and C_c. The commutating capacitor is charged in the reverse direction until it reaches supply voltage, V_d.

Interval E. The load current I_L continues to flow though L_c, until all the trapped magnetic field energy is converted to electrostatic energy is charging the capacitor. In the modified McMurray circuit (diodes D_{1A} and D_{2A} are across the auxiliary SCRs), the excess inductive energy is returned to the bus, thus reducing the commutating capacitor voltage. The overcharge across the commutating capacitor would appear as a forward voltage on the auxiliary SCRs, which requires a higher rated unit than would be normally used. Addition of diodes D_{1A}, D_{2A} and resistor R provide a path for this excess energy to be returned to the dc bus.

In the above circuit the turn-off time of the main SCRs must be lower than t_o as shown in Figure 8.19. The design of the McMurray circuit is given in the next section. The commutating capability of the McMurray circuit is a function of the load. As the load current increases the capacitor voltage increases and hence the capability to commutate higher current. Further, the commutation losses are lower at lower values of load current.

8.16 Design of McMurray Inverter Circuit

Selection of Optimum Commutating Capacitance and Inductance

The commutating current pulses could be of various magnitudes and widths to successfully commutate SCRs in the McMurray inverter as shown below. From Figure 8.20 it is clear that each current pulse would successfully commutate the SCRs if the turn-off time of the SCRs are below t_o shown in the figure. It has been shown by McMurray that the best efficiency of the commutating circuit is achieved when the peak of commutating pulse current is 1.5 times the load current. In our design criterion $I_c/I_L = 1.5$ would be used.

Calculation of Inductors and Capacitors for the McMurray Commutation Circuit

The commutating current pulse is shown in Figure 8.21. This current pulse is obtained when SCR1A is triggered to turn off SCR1. From

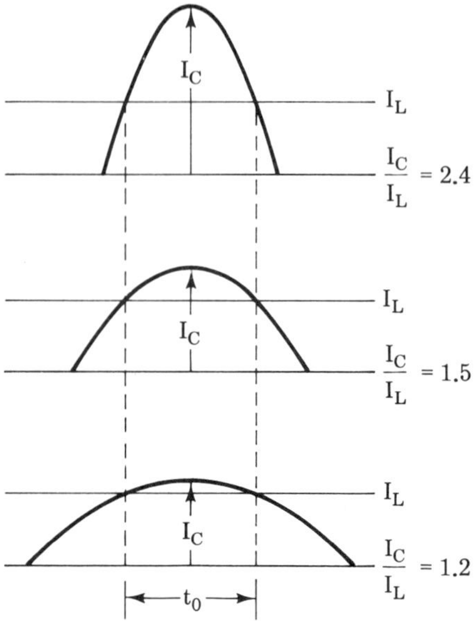

Figure 8.20. Commutating current pulses.

Figures 8.18 and 8.21:

V_c = voltage across commutating capacitor

I_c = peak value of commutating current

C_c = commutating capacitor

L_c = commutating reactor

I_L = load current

$\omega = 2\pi f$, where $f = \frac{1}{2t}$ = frequency of oscillation of the commutating pulse.

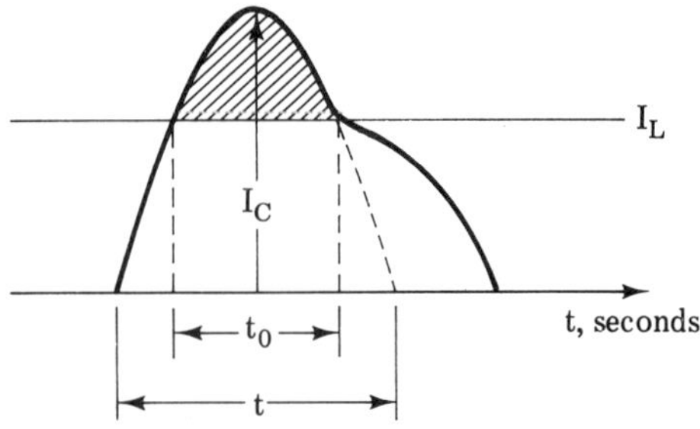

Figure 8.21. Commutating current pulse.

For minimum energy loss in the commutation circuit, $I_c = 1.5\, I_L$. For the commutation circuit:

$$I_c = \frac{V_c}{\omega L_c} = \frac{V_c}{2\pi f\, L_c} = \frac{V_c}{2\pi \cdot \dfrac{1}{2\pi\sqrt{L_c C_c}} \cdot L_c} = V_c \sqrt{\frac{C_c}{L_c}} \quad (8.28)$$

$$f = \frac{1}{2\pi\sqrt{L_c C_c}} \quad (8.29)$$

$$t = \pi\sqrt{L_c C_c} \quad (8.30)$$

For successful turning-off of the main SCRs, the commutation circuit must satisfy the following:

1. L_c and C_c should be chosen such that $t_o \geq$ turn-off time of SCR1.
2. V_c, L_c and C_c should be selected such that $I_c = 1.5\, I_L$.

EXAMPLE 8.4

In the McMurray inverter circuit (Figure 8.18a), calculate the values of L_c and C_c for the following conditions:

1. The maximum load current, $I_L = 100A$,
2. Turn-off time t_{off} of SCR1 is 40 μsec.
3. Input dc voltage, $V_d = 300V$.

SOLUTION

For minimum energy loss in the commutation circuit, $I_c = 1.5\, I_L = 1.5 \times 100 = 150A$. Let us limit the capacitor voltage V_c to 450 volts.

$$\therefore I_c = 150A = V_c \sqrt{\frac{C_c}{L_c}} = 450 \sqrt{\frac{C_c}{L_c}}$$

or

$$\sqrt{\frac{C_c}{L_c}} = \frac{150}{450} = \frac{1}{3} \quad (8.31)$$

From Figure 8.21, $t_o = 0.535\, t$. In the given problem $t_{off} = 40$ μsec. Taking 50% tolerance, $t_o = 60$ μsec

$$\therefore t = \frac{60\mu sec}{0.535} = 112\mu sec$$

From Eq. 8.30 $t = \pi\sqrt{L_c C_c}$

$$\therefore \sqrt{L_c C_c} = \frac{112 \times 10^{-6} sec}{\pi} = 35.65 \times 10^{-6} sec \quad (8.32)$$

From Eq. 8.31

$$\sqrt{\frac{L_c}{C_c}} = 3 \tag{8.33}$$

Solving Equations 8.32 and 8.33,

$$L_c = 107\mu H$$

$$C_c = 11.88\mu F$$

8.17 dV/dt and dI/dt Calculation of the Main and Commutating SCRs

The rate of change of voltage across the thyristors and the rate of change of current through the thyristors have to be limited to the maximum values allowed according to the manufacturer's data for the device. The voltage seen by the auxiliary SCR is the commutating capacitor voltage.

Commutating SCR: dV/dt Calculation

The commutating capacitor is charged to $V_c = 450$ volts. The point A (Figure 8.18a) is at dc supply voltage of 300V through SCR1. The point B must be at negative 150V. Hence when SCR1A is triggered, the voltage change across SCR2A is 450 volts through L_3, SCR1A and L_4. Voltage oscillation across SCR2A is given by $V = V_c \sin \omega t$, where $V_c = 450$ volts,

$$\omega = \frac{1}{\sqrt{LC}} \text{ and } L = L_3 + L_4$$

$$C = C_{s2}$$

$$\therefore \frac{dV}{dt} = \omega V_c \cos \omega t$$

Hence,

$$\left.\frac{dV}{dt}\right|_{max} = \omega V_c - \frac{V_c}{\sqrt{(L_3 + L_4)C_{s2}}} \tag{8.34}$$

Similarly, dV/dt across SCR1A could be calculated when SCR2A is triggered.

EXAMPLE 8.5. *dV/dt Calculation*

Assume maximum dV/dt of SCR1A is 200V/μsec, $V_c = 450V$. Choose $C_{s2} = .01\ \mu F$ (increase in value of C_{s2} increases snubber losses).

From Eq. 8.34,

$$\left.\frac{dV}{dt}\right|_{max} = \frac{200V}{\mu sec} = \frac{450}{\sqrt{(L_3 + L_4)^{-6}.01 \times 10^{-6}}}$$

$$\therefore \sqrt{(L_3 + L_4).01} = \frac{450}{200}$$

or
$$(L_3 + L_4).01 = 5.06$$

or
$$L_3 + L_4 = 506\mu H$$

So, to limit dV/dt to 200V/μsec, L_3 and L_4 should be 253 μH each.

Commutating SCR dI/dt Calculation

When SCR1A is triggered to turn off SCR1, the rate of change of current through SCR1A is given by

$$L\frac{dI}{dt} = V_c = 450V$$

where
$$L = L_1 + L_3 + L_c$$

$$\therefore \frac{dI}{dt} = \frac{450}{L_1 + L_3 + L_c}$$

EXAMPLE 8.6

Assume dI/dt of the auxiliary SCR1A is 20A/μsec. Then,

$$L_1 + L_3 + L_c = \frac{450}{20\mu A/\mu sec} = 22.5\mu H$$

Hence, the auxiliary SCR's dI/dt rating can be chosen to be low. Snubber circuit resistance should be chosen as

$$R_{s1} = \sqrt{\frac{(L_3 + L_4)}{C_{s2}}}\ ohm$$

The value of reactors L_3 and L_4 are required for dV/dt limitation is much higher than for limiting dI/dt. Hence saturable reactors would be used for L_3 and L_4 which would allow a high initial value for dV/dt limitation.

8.18 The Snubber Circuit

The R-C circuit connected across each SCR is known as the snubber circuit. The snubber circuit is used to limit the rate of rise of voltage (dV/dt) across the SCR during switching.

8.19 Main SCR Calculation

The voltage rating of the main SCR is the dc supply voltage.

dI/dt Calculation

When SCR1 is turned on, the rate of change of current through the device is given by

$$V_d = L\frac{dI}{dt}, \text{ where } L = L_1 + L_2$$

$$\therefore \frac{dI}{dt} = \frac{V_d}{L_1 + L_2} = \frac{300}{L_1 + L_2}$$

If dI/dt is 100A/μsec, then,

$$L_1 + L_2 = \frac{300}{100}\,\mu H = 3\mu H$$

Hence, dI/dt of the SCR chosen could be quite low by choosing higher values of inductances.

dV/dt Calculation

When SCR1 is turning on, the change of voltage V_d across SCR2 is through L_1, R_{s1}, C_{s1}, L_2 and SCR1. The oscillation voltage V is given by

$$V = V_d \sin \omega t$$

or

$$\frac{dV}{dt} = \omega \cdot V_d \cos \omega t$$

or

$$\left.\frac{dV}{dt}\right|_{max} = \omega V_d \text{ where } V_d = 300 \text{ volts and, } \omega = \frac{1}{\sqrt{LC}}.$$

But

$$L = L_1 + L_2, \text{ and } C = C_{s1}$$

$$\therefore \left.\frac{dV}{dt}\right|_{max} = \frac{300}{(L_1 + L_2)C_{s1}}$$

EXAMPLE 8.7

Assume dV/dt of the main SCR is 200V/μsec.

Lot $C_{s1} = .25\mu F$

Then $\sqrt{(L_1 + L_2)} = \dfrac{300}{\sqrt{C_{s1}}\,200}$

$$\therefore (L_1 + L_2) = \frac{2.25}{.25}\ \mu H = 9\mu H$$

$$R_{s1} = \sqrt{\frac{L_1 + L_2}{C_{s1}}}\ \text{ohms} = \sqrt{\frac{9\mu H}{.25}} = 6\ \text{ohms}$$

8.20 Snubber Circuit Losses

The loss in the snubber circuit is a function of the capacitor value, the voltage across the capacitor, and the switching frequency.

Main SCR Snubber Circuit Loss for 60° Chopping

$$P_{max} = \text{snubber circuit loss}$$

$$= \frac{1}{2} C_{s1} V_d^2 f$$

where $f = 2F_{chopping} \cdot \dfrac{1}{3} + 2f_m$

$F_{chopping}$ = chopping frequency

= 600Hz, $C_{s1} = 0.25\mu F$ = snubber capacitor

f_m = inverter frequency, V_d = 300 volts = input dc voltage

f_m = 60Hz

$$P_{max} = \frac{1}{2} \times 0.25 \times 10^{-6} \times 300^2 \times 2\left(600 \times \frac{1}{3} + 60\right)$$

$$= 5.85\ \text{watts}$$

Commutating SCR Snubber Circuit Loss for 60° Chopping

$$P_{max} = \text{snubber circuit loss} = \frac{1}{2} C_{s2} V_c^2 \cdot f$$

where $f = 2F_{chopping} \cdot \dfrac{1}{3} + 2f_m$

$F_{chopping}$ = chopping frequency = 600Hz

$$f_m = \text{inverter output frequency} = 60\text{Hz}$$

$$V_c = \text{commutating capacitor voltage} = 450\text{V}$$

$$C_{s2} = .01\mu\text{F} = \text{snubber capacitor}$$

$$P_{max} = \frac{1}{2} \times .01 \times 10^{-6} \times 450^2 \times 2 \left(600 \times \frac{1}{3} + 60\right)$$

$$= 0.52 \text{ watt}$$

8.21 McMurray-Bedford Circuit

Figure 8.22 shows one phase of a three-phase inverter using the McMurray-Bedford circuit. The two halves of the commutating inductance are tightly coupled. The period of the ac output of the inverter is much greater than the turn-off time of the SCRs. The commutation process is initiated when the nonconducting SCR is triggered on to turn off the conducting SCR. One half-cycle of operation, during the commutation process with a lagging load, is shown in Figure 8.23. The five distinct intervals during commutation are explained below and the corresponding waveforms are shown in Figure 8.23.

Interval 1. SCR1 is conducting current to the load drawing power from the positive side of the dc supply. With an inductive load, the rate of change of current through inductor L1 is negligible, the voltage across inductance L1 is very small compared to E_d. Hence the terminal

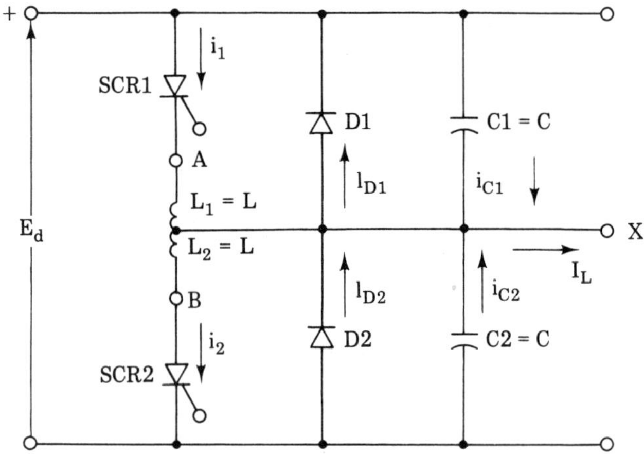

Figure 8.22. One phase of a three-phase inverter employing the McMurray-Bedford commutation method.

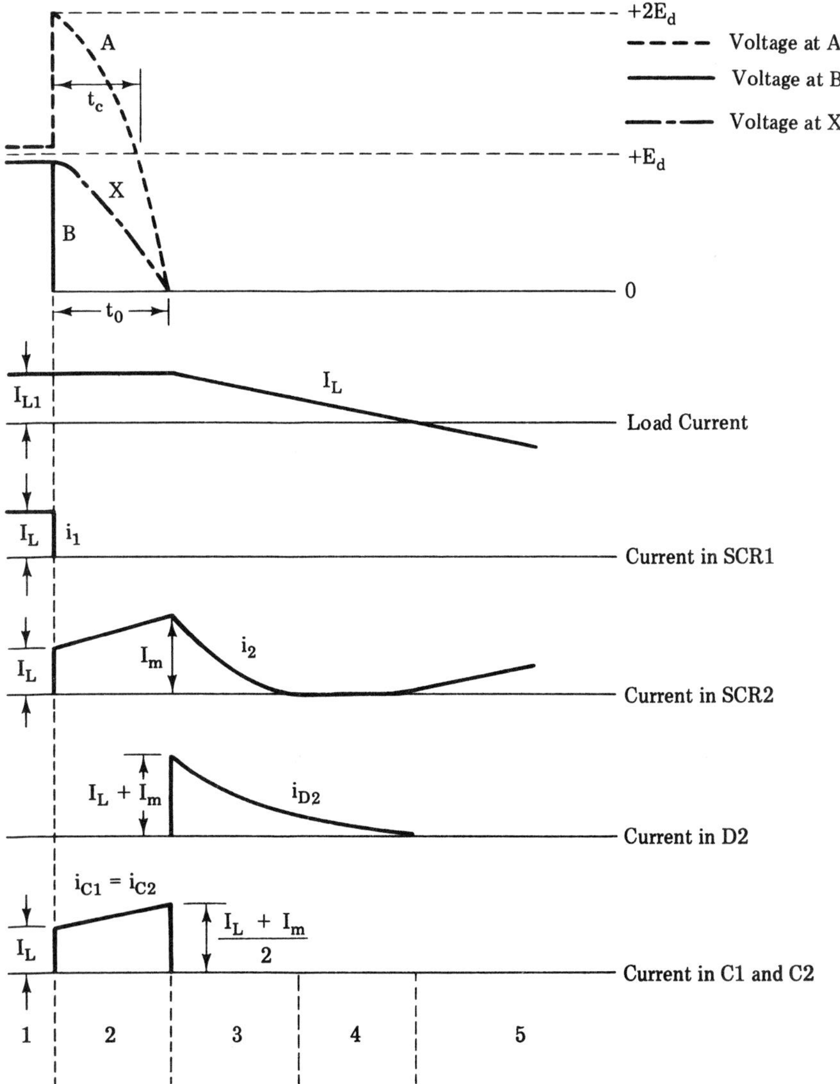

Figure 8.23. Voltage and current waveforms in the McMurray-Bedford circuit with lagging load.

X of the load is at the positive of the dc supply voltage E_d. The capacitor C1 is charged to zero voltage, and C2 to E_d. The load current is equal to I_L.

Interval 2. To turn off SCR1, thyristor SCR2 is triggered; at this instant, point B potential falls to negative of the supply. Since the voltage across capacitors C1 and C2 cannot change instantaneously, a voltage E_d must appear across inductance L_2. The same voltage is induced across $L_1(E_d)$. Thus the potential of point A is $2E_d$, which is across SCR1 and turns it off. The load current I_L, which was flowing through SCR1 and L_1, now transfers to winding L_2 and SCR2 in order to maintain the energy stored in the inductance. The constant load current I_L, is supplied by the capacitors C1 and C2. These initial conditions and waveforms are shown in Figure 8.23. The waveform is part of a sinusoid of period $2\pi\sqrt{2LC}$. The duration of the commutating interval t_o is about ¼ cycle. The time t_c available for the SCR1 to turn off is even shorter. During the commutating interval, the voltage across capacitor C2 is also across L_2. As the voltage across capacitor C2 discharges, the current i_2 increases from I_L to I_m and the voltage across C2 goes to zero.

Interval 3. The onset of interval 3 is delayed until point X reaches the negative of the supply. The capacitor C1 is now charged to the input voltage E_d. The feedback diode D2 starts to conduct and connects the point X to the negative of the supply and supplies the load current I_L. The trapped energy in L_2 dissipates by circulating current through L_2, SCR2 and D2. By proper design of the circuit the trapped energy in L_2 may be fed back to the dc supply or it may be quickly dissipated by putting a ıesistance in series with diode D2.

Interval 4. The turn-off process of SCR1 is complete when the trapped energy and the circulating current have been reduced to zero. The inductive energy of the load keeps diode D2 in conduction. After i_2 goes to zero, SCR2 is reverse biased and is turned off. The current i_{D2} finally decays to zero and the load current I_L finally also goes to zero and reverses at the end of this interval.

Interval 5. SCR2 has to be retriggered at this point, if the original gate pulse has been removed after initial triggering during interval 2. The load current in the reverse direction is now carried by SCR2 as shown in Figure 8.23. The transfer of load current from SCR1 to SCR2 is now complete.

There are several modifications of the McMurray-Bedford circuit to return the trapped energy in the inductance to the dc supply.

8.22 Pulse Width Modulated Inverter (PWM)

The voltage applied to the terminals of an ac induction motor must be varied directly with frequency in order to maintain constant flux in the machine. The voltage must be held within a relatively narrow range at any one frequency in order to provide proper excitation. Over-excitation results in saturation which produces high peak armature current. Under-excitation reduces the available torque which the motor can produce since torque varies as the square of the flux.

A number of different methods have been used for controlling the output voltage of an inverter. The dc input voltage or the ac output voltage can be directly controlled, or when several inverters are used, their output voltage can be phase shifted and added. All of these methods require additional controlled power conversion stages, and the first one also requires a separate commutation voltage supply.

In the pulse width modulation (PWM) scheme, the voltage output of the inverter is varied by using time ratio control inside the inverter. The result is a savings in power components at a sacrifice of increased control complexity. The inverter control must now synchronize the SCR firing and commutation to produce the desired three-phase output voltage and the desired three-phase output frequency.

8.23 Different Chopping Techniques

Several chopping methods and frequencies are listed below to obtain pulse width modulation for voltage control.

1. Variable chopping frequency synchronized to fundamental frequency.
   Advantages: Simple control is required with good volts/Hz linearity.
   Disadvantages: Results in high current ripple at lower fundamental frequencies because of low chopping frequency; therefore, severely limits motor speed range.
2. Constant chopping frequency—not synchronized to fundamental frequency.
   Advantages: Keeps current ripple within acceptable limits at low fundamental frequencies.
   Disadvantages: Results in dc or lower frequency components due to unequal volt-sec per half cycle at certain fundamental to chopping frequency ratios.
3. Constant chopping frequency, but chopping in each phase only in the middle 60° section of each 180° half-wave.
   Advantages: Reduces current ripple due to free-wheeling effect during "off" periods. Results in one-third the number of commutations as the first two approaches mentioned above, since chopping occurs only in one phase at a time. Control components

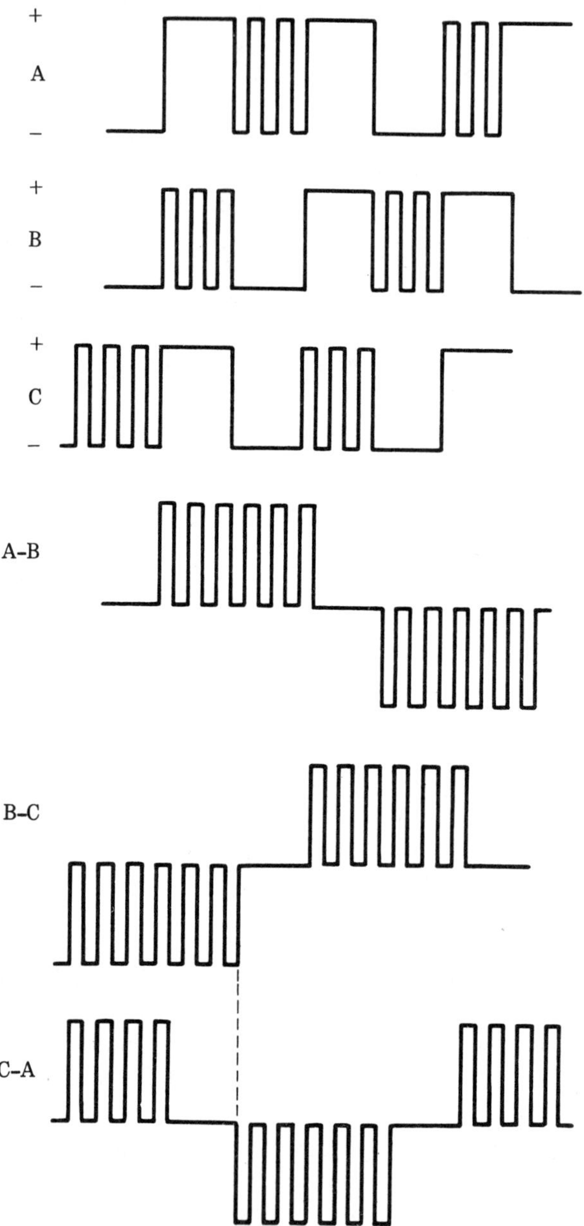

Figure 8.24. Switching diagram when chopping.

can be reduced since one chopping logic system can be "time shared" between the three phases. Chopping is not affected by a phase lag of current from voltage up to 60°.

Disadvantages: Cannot obtain zero voltage unless chopping frequency is reduced. Contains steps in volts/Hz characteristic which must be regulated out. Limited to square wave output only—sine-wave shaping not possible.

8.24 Sixty-Degree Chopping

The 60° approach provides satisfactory operation with the minimum number of commutations and with relatively simple and inexpensive control. The 60° chopping method is shown in Figure 8.24, which shows how this chopping scheme can still produce a completely controlled phase-to-phase voltage. The control sequence, which is repeated, consists of chopping in one phase for 60°, changing polarity of a second phase, and chopping in the third phase for 60°, etc. It can be seen that one set of chopping logic can be used if its output is distributed to the correct phase control at each instant in time.

The actual time ratio control, the on versus off time for any fundamental frequency, is generated in a linear circuit. A voltage reference level is compared with a sawtooth voltage generated at the desired chopping frequency, as show in Figure 8.25. Depending on the magnitude of the reference voltage in comparison with the sawtooth voltage, the on versus off time is varied from practially all off to all on.

The actual chopping frequency is a compromise based on considerations of motor heating, commutation losses, and SCR switching characteristics.

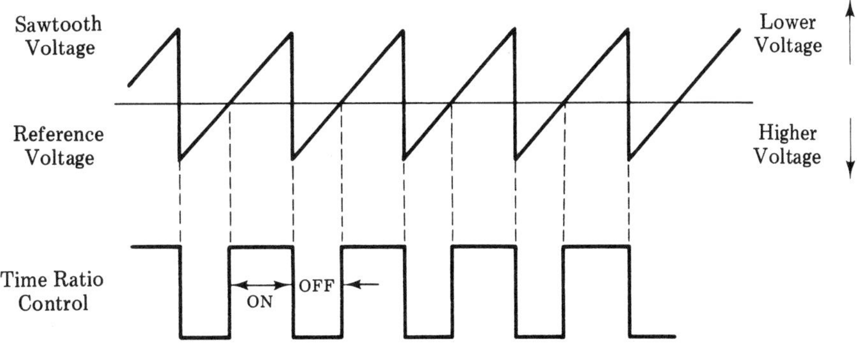

Figure 8.25. Time ratio voltage generation.

8.25 Commutation Circuit Design

The losses in the inverter are due to:

1. Loss in the commutation circuit.
2. Loss in the snubber circuit.
3. Loss in the SCRs and diodes.
4. Loss due to reverse leakage current.

Commutation and snubber losses are proportional to frequency of operation of the inverter. To increase the efficiency of the inverter, the losses in the commutation and snubber circuits should be minimized.

chapter NINE

Control of DC and AC Motors

9.1 Propulsion of DC Motors

The speed of a dc motor is controlled by varying the voltage to its armature or field. By varying the voltage to the field, one can only run the motor at speeds higher than the base speed at the expense of falling torque. The important aspect of the speed control of a dc motor is the armature voltage control method. By varying the voltage to the armature of the dc motor, the speed of the motor could be varied from almost zero speed to its rated speed at constant torque.

The block diagram of a typical speed control system using a three-phase full-wave controlled rectifier is shown in Figure 9.1. The control of voltage to the armature and the field enables speed control below and above base speed of the motor.

The operating characteristics of the separately excited dc motor are shown in Figure 9.2. As seen from Figure 9.2, the air gap flux remains constant up to rated speed, hence the torque is maintained constant at variable speed up to rated speed (armature voltage control). Beyond rated speed, air gap flux drops, hence speed increases with decreasing torque (field control).

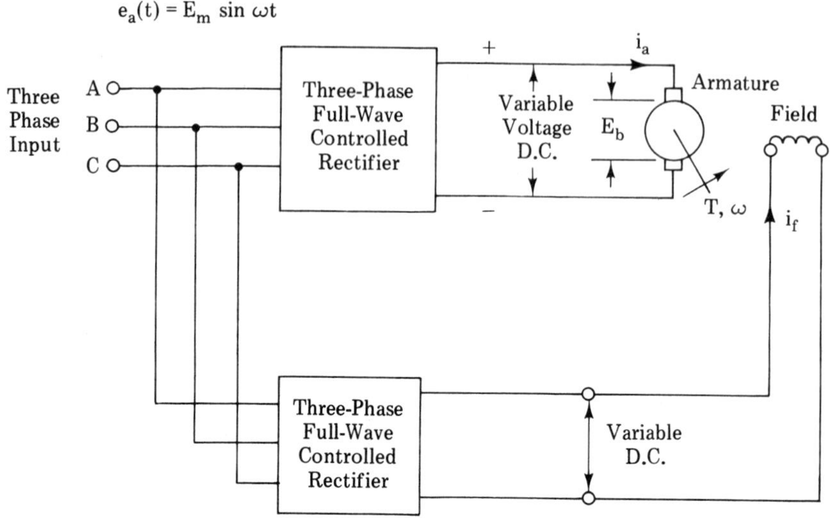

Figure 9.1. Block diagram of a dc motor speed control system.

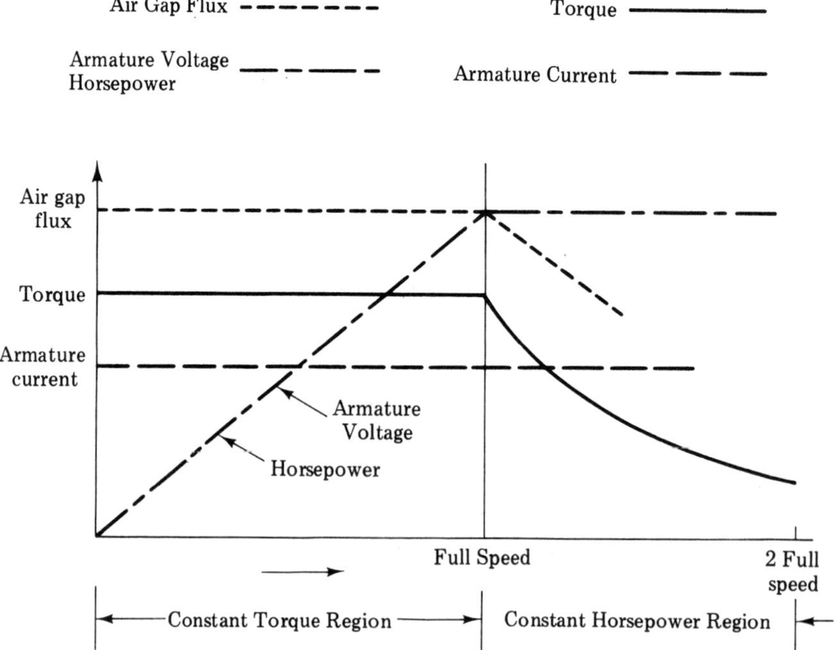

Figure 9.2. Operating characteristics of a separately excited dc motor.

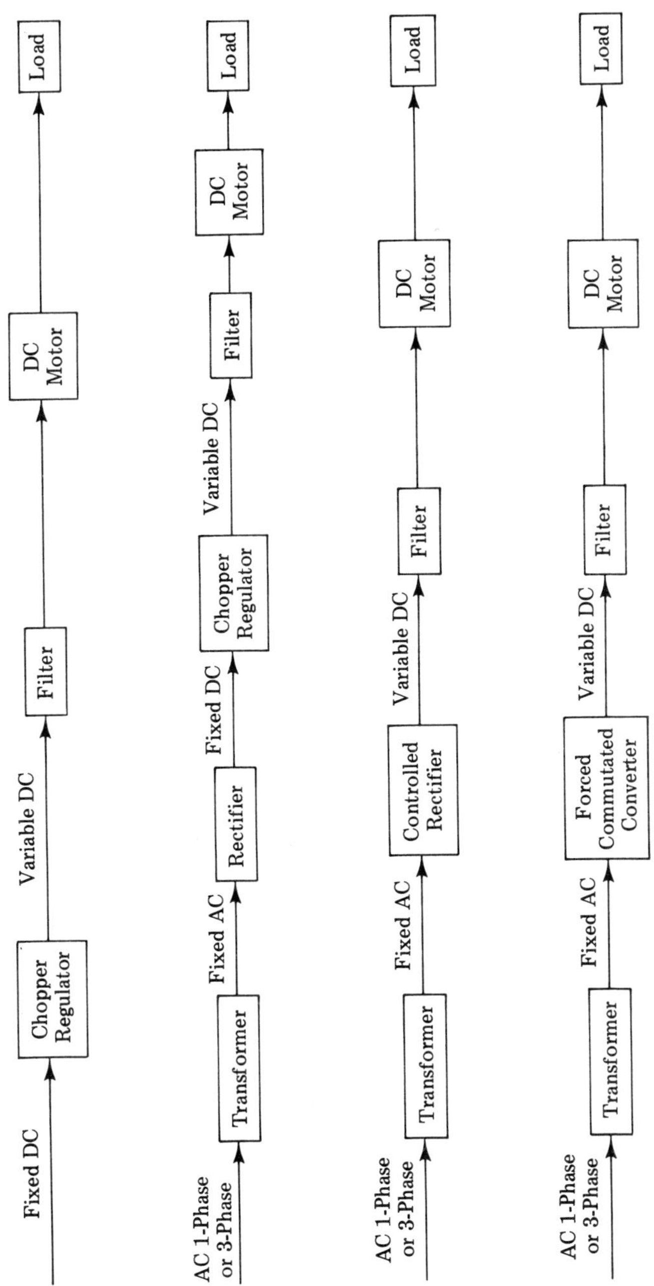

Figure 9.3. Schemes of dc motor speed control.

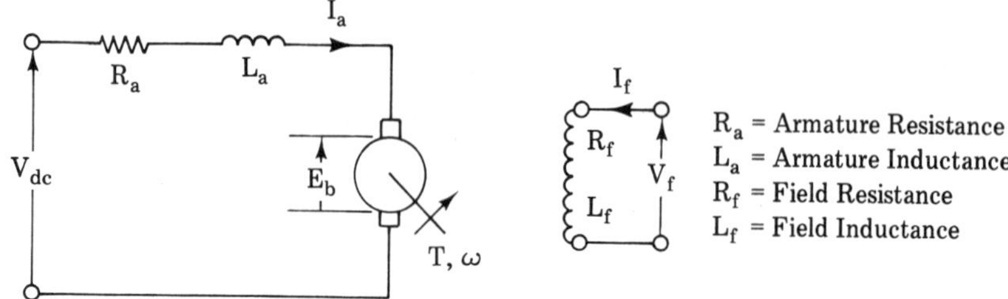

Figure 9.4. Model of a dc motor.

9.2 Various Schemes of DC Motor Speed Control

The speed of a dc motor has to be controlled from an ac or dc source. This section gives different schemes of the dc motor control as shown in Figure 9.3.

9.3 Model of a DC Motor

The model of a dc motor is shown in Figure 9.4.

The basic speed and torque equations of a dc machine under steady-state are given by the following:

$$\omega = \frac{E_b}{K_b\phi} \tag{9.1}$$

$$T = K_t\phi I_a \tag{9.2}$$

In the above two equations:

ω = speed of the motor in radians/sec

E_b = back emf of the motor

K_b = motor speed constant

ϕ = flux in the machine

T = torque developed by the motor

K_t = torque constant

I_a = armature current

Also, $$E_b = V_{dc} - I_aR_a \tag{9.3}$$

where V_{dc} = supply voltage to the motor

R_a = armature resistance

It is seen from Equation 9.1 that the speed, ω, of the motor is a function of the back emf E_b, motor constant K_b, and flux ϕ. From

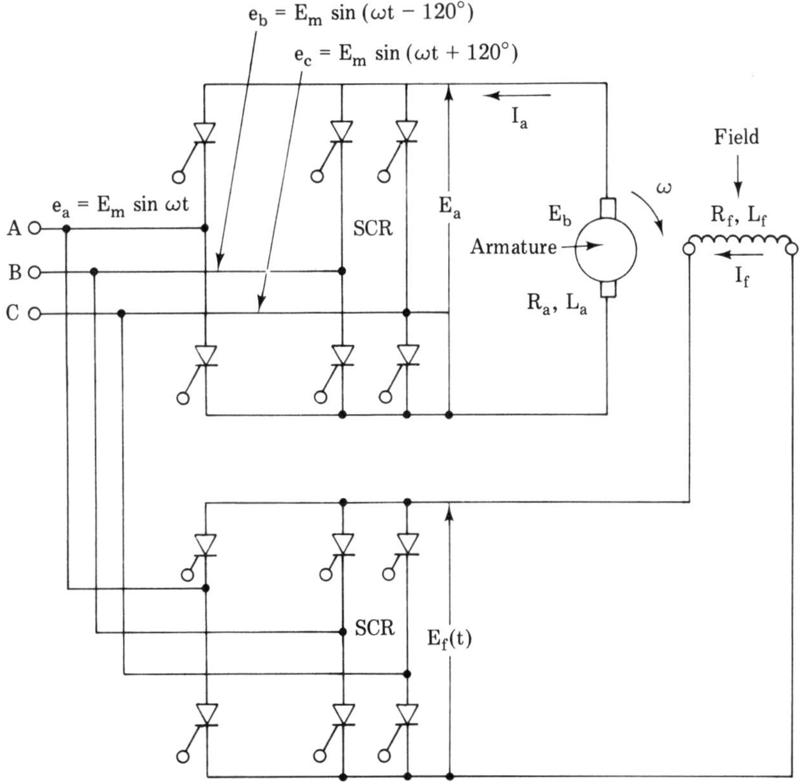

Figure 9.5. Circuit diagram of a dc motor control system.

Equation 9.2 the torque is a function of K_t, ϕ, and I_a. So the speed of the dc motor can be varied from zero to rated speed by varying the voltage V_{dc} supplied through a phase-controlled rectifier. The torque could be maintained constant at the rated value by maintaining constant flux in the machine. Flux is maintained constant by constant field current. For running in the constant horsepower region, the supply voltage V_{dc} is maintained at rated value and the field voltage is decreased to reduce the field flux. From Equation 9.2 it is clear that the torque in the machine could fall in the field control region as seen in Figure 9.2.

9.4 Separately Excited DC Motor Running from a Three-Phase Full-Wave Controlled-Rectifier Circuit

The armature and the field circuit in Figure 9.5 is supplied through a 3-ϕ full-wave SCR circuit. Suppose sinusoidal voltages $e_a(t)$, $e_b(t)$, and $e_c(t)$ were applied to the terminals A, B, and C. There flows current

I_a in the armature and $I_f(t)$ in the field. Thus, the speed voltage $KI_f\omega$ proportional to speed and the field current I_f may be induced in the armature where K is a constant of the machine and the effect of saturation is neglected.

9.5 Output Voltage in the Field Current

If harmonics of the output voltage are neglected, then for three-phase full-wave rectification with delay angle α (assuming continuous conduction), the field voltage E_f is given by

$$E_f = \frac{3E_m}{\pi} \cos \alpha \tag{9.4}$$

$$\therefore \text{ field current } I_f = \frac{3E_m \cos \alpha}{\pi R_f} \tag{9.5}$$

The lowest harmonic would be the sixth harmonic which could be neglected, because of high inductance in the field circuit.

9.6 Output Voltage in the Armature Circuit

Since SCR is used in the phase-controlled rectifier for the armature circuit, the output voltage could be varied (depending on phase-delayed control) from the rectifier and hence, the voltage to the motor.

Assuming continuous conduction for all delay angles, and neglecting harmonics, the armature voltage is given by

$$E_a = \frac{3E_m}{\pi} \cos \alpha \tag{9.6}$$

The torque developed by the motor is given by

$$T_m = K_m I_f I_a \tag{9.7}$$

The armature current, I_a, is given by

$$I_a = \frac{E_a - K_f I_f \omega}{R_a} \tag{9.8}$$

Speed of motor W is given by

$$\omega = \frac{E_a - I_a R_a}{K_f I_f} \tag{9.9}$$

Substituting the value of I_a from Equation 9.8 in Equation 9.7,

$$T_m = K_m I_f \frac{E_a - K_f I_f \omega}{R_a}$$

$$= K_m I_f \frac{3E_m \cos \alpha/\pi - K_f I_f \omega}{R_a} \quad (9.10)$$

If torque is zero, i.e., under no-load condition, Equation 9.10 reduces to

$$\frac{3E_m}{\pi} \cos \alpha = K_f I_f \omega_{no\ load}$$

or

$$\omega_{no\ load} = \frac{3E_m \cos \alpha}{\pi K_{f1}} \quad (9.11)$$

if I_f is constant.

Equation 9.11 shows that no-load speed of the motor is a function of cos α where α is the delay angle of triggering of the SCRs in the armature. In the constant torque region, the field current is maintained constant at its rated value and the armature voltage is varied for control of speed at constant torque. For constant horsepower operation, the field current is reduced to run the motor above rated speed.

9.7 Separately Excited DC Motor Running from a Single-Phase Full-Wave Controlled-Rectifier Circuit

The speed of a dc motor could also be controlled from a single-phase full-wave circuit. Several different modifications of the bridge circuit have been discussed in Chapter 4. Two typical bridge configurations used in the control of dc motors are described in the next two sections.

The back emf of the motor is given by

$$e_b = K_b \phi n \quad (9.12)$$

where e_b is the instantaneous back emf and n is the instantaneous speed, under steady-state condition,

$$E_b = K_b \phi N \quad (9.13)$$

where

E_b = average voltage

N = steady-state speed

ϕ = flux in the machine

The torque developed by the motor is given by

$$t = K_b \phi i_a \quad (9.14)$$

where i_a is the instantaneous current into the armature of the motor and t is the torque. The average developed torque is

$$T = K_b \phi I_a \tag{9.15}$$

where I_a is average dc current into the armature of the motor.

The voltage equation of the armature circuit is given by

$$e_a = R_a i_a + L_a \frac{d_{ia}}{dt} + e_b \tag{9.16}$$

The average armature voltage is given by

$$E_a = R_a I_a + E_b \tag{9.17}$$

The steady-state speed of the motor is given by

$$N = \frac{E_a - I_a R_a}{K_b \phi} \tag{9.18}$$

9.8 Speed Control of DC Motors Using Two Thyristors and Two Diodes in a Bridge Connection

A speed control scheme using a free-wheeling diode is shown in Figure 9.6. The free-wheeling diode D3 is necessary to allow circulating current in the motor during the off period of the thyristors. A sufficiently large inductance in series with the armature and a free-wheeling diode prevent discontinuous current conduction in the armature of the motor. Discontinous current conduction causes severe torque and speed ripple in the motor.

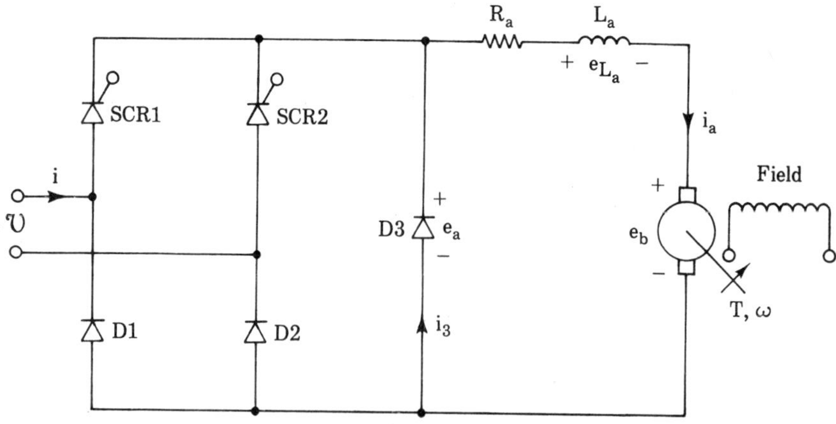

Figure 9.6. Speed control of a separately excited dc motor by a single-phase full-wave bridge circuit with two thyristors and two diodes.

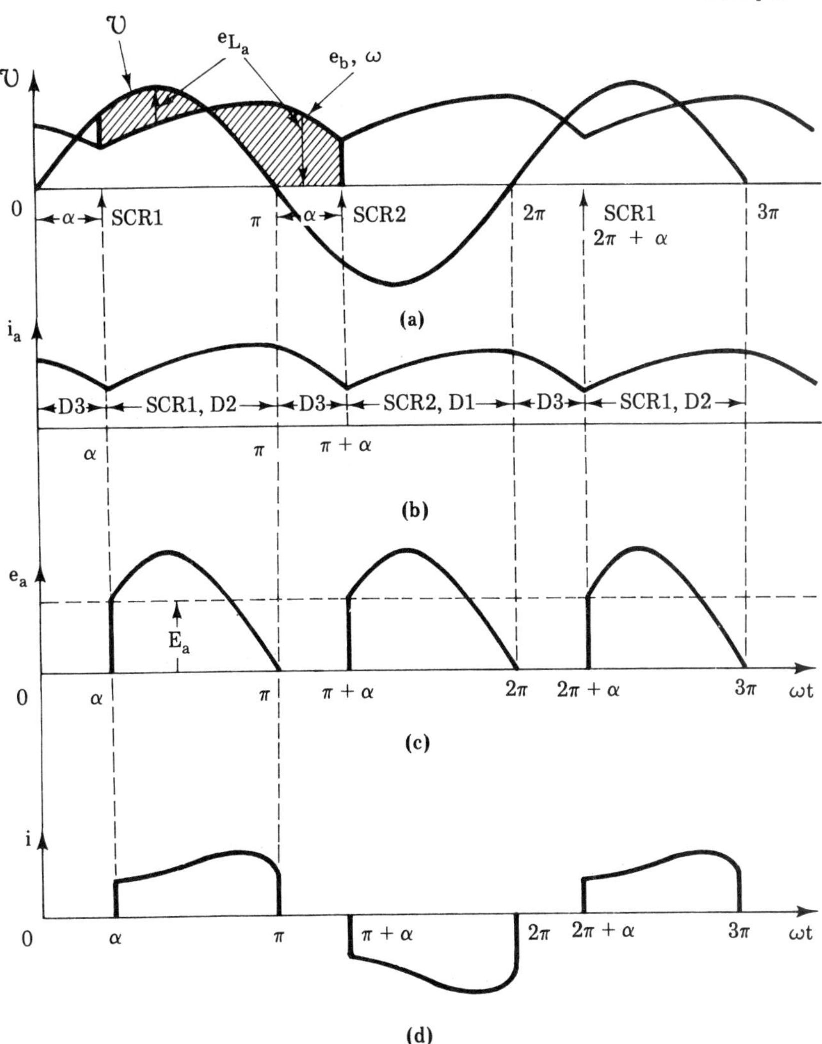

Figure 9.7. Speed control of a separately excited dc motor by a single-phase full-wave circuit. Voltage and current waveforms for continuous motor current: (a) input voltage; (b) armature current; (c) output voltage; (d) input current.

The motor voltage and current waveforms for continuous current conduction are shown in Figure 9.7.

In Figure 9.7(a), the thyristors are fired at a delay angle α. SCR1 starts conducting at α and turns off at π. At π free-wheeling diode D3 starts to conduct and carries the armature current till π + α. The armature current waveform is shown in Figure 9.7(b). Figure 9.7(c) shows the output voltage, e_a, of the controlled rectifier. The output voltage, e_a, is zero when the free-wheeling diode is conducting. The input current, i, is shown in Figure 9.7(d). It is seen that the input current is not continuous.

The armature voltage equations are as follows:

$$e_a = v = R_a i_a + L_a \frac{d_{ia}}{dt} + e_b, \ \alpha < \omega t < \pi \tag{9.19}$$

$$e_a = 0 = R_a i_a + L_a \frac{d_{ia}}{dt} + e_b, \ \pi < \omega t < \pi + \alpha \tag{9.20}$$

The average (dc) value of the motor terminal voltage is

$$E_a = \frac{1}{\pi}\int_{\alpha}^{\pi} V_m \sin \omega t \, d(\omega t) = \frac{V_m}{\pi}(1 + \cos \alpha) \tag{9.21}$$

The steady-state speed equation is given by

$$N = \frac{E_a - I_a R_a}{K_b \phi} = \frac{V_m(1 + \cos \alpha)}{\pi K_b \phi} - \frac{R_a T}{(K_b \phi)^2} \tag{9.22}$$

where $I_a = \dfrac{T}{K_b \phi}$, from Equation 9.15.

The no-load speed of the motor is given by

$$N_{N \cdot L} = \frac{V_m(1 + \cos \alpha)}{\pi K_b \phi} \quad \text{where } T = 0$$

9.9 Speed Control of DC Motors Using Four Thyristors in a Bridge Connection

The speed control scheme is shown in Figure 9.8. With sufficient inductance (L_a) in the armature, this type of circuit is capable of operating in the regenerating mode and also the current conduction is continuous. The waveforms under continuous current conduction are shown in Figure 9.9. Figure 9.9(a) shows the input voltage and the shaded area shows the voltage across the inductance. The triggering points of the thyristors are also shown in the figure. SCR1 and SCR3 are fired at a delay angle of α. These two thyristors continue to conduct until $\pi + \alpha$, when SCR2 and SCR4 are triggered. The armature current waveform is shown in Figure 9.9(b). The output voltage, e_a of the controlled rectifier is shown in Figure 9.9(c). The average value of the output voltage is also shown. The input current i is shown in Figure 9.9(d). The input current is continuous in this case.

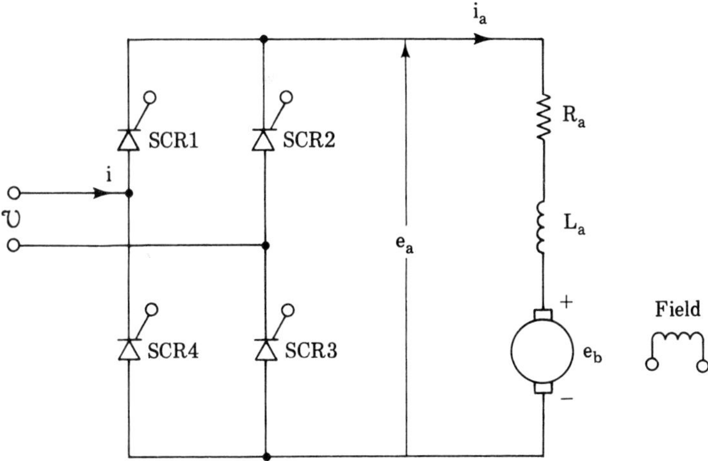

Figure 9.8. Speed control of a separately excited dc motor by a single-phase full-wave bridge with four thyristors.

The average armature voltage is given by

$$E_a = \frac{1}{\pi}\int_{\alpha}^{\pi+\alpha} V_m \sin\theta \, d\theta = \frac{2V_m}{\pi}\cos\alpha \tag{9.23}$$

The steady-state speed equation is given by

$$N = \frac{E_a - I_a R_a}{K_b\phi} = \frac{2V_m \cos\alpha}{\pi K_b \phi} - \frac{R_a T}{(K_b\phi)^2} \tag{9.24}$$

where $I_a = \dfrac{T}{K_b\phi}$, from Equation 9.15.

The no-load speed of the motor is given by

$$N_{N \cdot L} = \frac{2V_m \cos\alpha}{\pi K_b \phi}, \text{ when } T = 0$$

Figure 9.10 shows the operation of the circuit of Figure 9.8 when operating in the inversion mode. If the delay angle α is greater than 90°, the average dc voltage E_a is negative. In this mode of operation, the dc machine acts as a generator and it can feed back power to the ac source. The inversion mode of operation is used for regenerative braking of the motor.

EXAMPLE 9.1

The speed of a 20HP, 460V, 1000 rpm separately excited dc motor is controlled by a single-phase full-wave bridge circuit as shown in Fig-

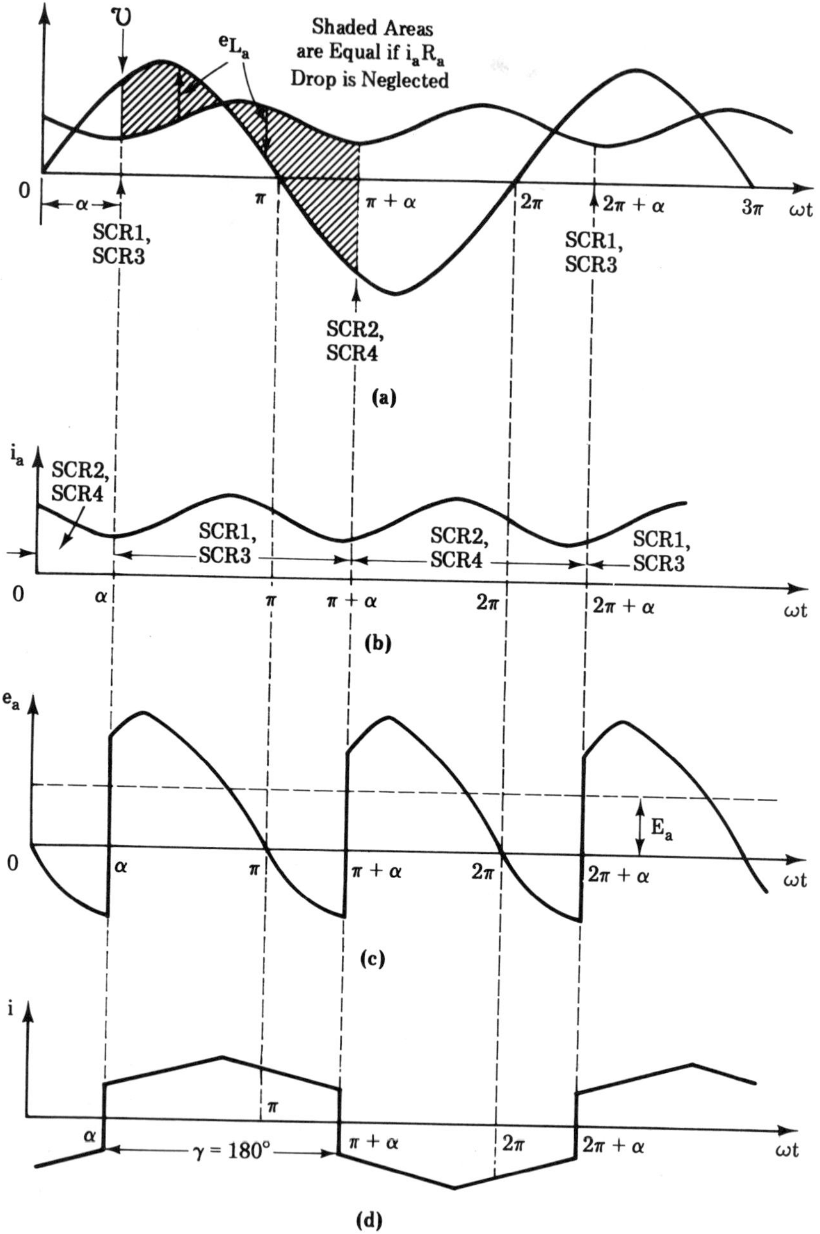

Figure 9.9. Speed control of a separately excited dc motor by a single-phase full-wave controlled rectifier. Voltage and current waveforms for continuous motor current (motor operation).

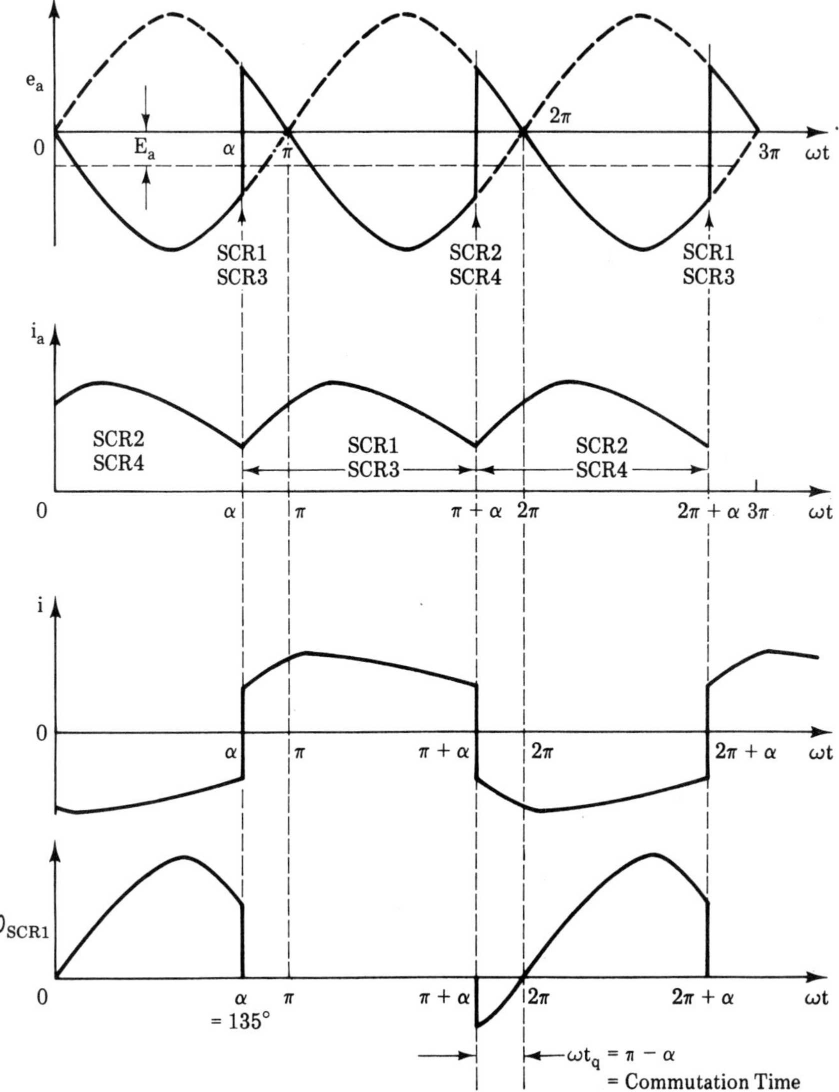

Figure 9.10. Waveforms for inversion operation.

ure 9.8. The rated motor armature current is 35A, and the armature resistance is .15Ω. The ac supply voltage is 480V. The motor back emf constant is $K_b\phi$ = .45V/rpm. The motor current is continuous and ripple free.

(a) Rectifier operation (dc machine running as a motor).

For a firing angle $\alpha = 60°$ and rated armature current, calculate
1. Torque
2. Speed
3. Supply power factor

SOLUTION

(1) $$K_b\phi = .45\text{V/rpm} = \frac{.45 \times 60}{2\pi}\text{ Vsec/rad} = \frac{4.297\text{V sec}}{\text{rad}}$$

$$T = K_b\phi I_a = 4.297 \times 35 = \mathit{150.4\ N\text{-}M}$$

(2) $$E_a = \frac{2V_m}{\pi}\cos\alpha = \frac{2 \times \sqrt{2} \times 480}{\pi}\cos 60° = 216.1\text{V}$$

$$E_b = E_a - I_aR_a = 216.1 - (35 \times .15) = 210.85\text{V}$$

$$\text{Speed } N = \frac{E_b}{K_b\phi} = \frac{210.85}{.45} = \mathit{468.6\ rpm}$$

(3) $$I_a = 35\text{A}$$

$$\text{Input volt-ampere} = VI = 480 \times 35 = 16800\text{VA}$$

$$\text{Real power} = E_aI_a = 216.1 \times 35 = 7563.5 \text{ watts}$$

$$\text{Power factor} = \frac{7563.5}{16800} = \mathit{.45}$$

(b) Inverter operation (regenerative).
The motor back emf polarity is reversed by reversing the field excitation. Calculate the following:
1. The firing angle to keep the motor current at its rated value.
2. The amount of power fed back to the supply.

SOLUTION

(1) At the instant of polarity reversal, the back emf is:

$$E_b = 210.85\text{V}$$

For the generator equation,

$$\begin{aligned} E_a &= E_b + I_aR_a \\ &= -210.85 + (35 \times .15) \\ &= -205.6\text{V} \end{aligned}$$

Also, $$E_a = \frac{2 \times \sqrt{2} \times 480}{\pi}\cos\alpha = -205.6$$

or $$\cos \alpha = .47576$$

$$\alpha = \mathit{118.4°}$$

(2) Power from the dc machine as a generator is

$$P_g = 210.85 \times 35 = 7379.75 \text{ watts}$$

Power loss in the armature resistance

$$= 35^2 \times .15 = 183.75 \text{ watts}$$

Power fed back to the supply $= 7379.75 - 183.75$
$= \mathit{7196\ watts}$

Power fed back to the supply may also be calculated as

$$E_a I_a = 205.6 \times 35 = \mathit{7196\ watts}$$

9.10 Problems of DC Motor Drives Operating From Phase-Controlled Rectifier

The following problems arise in the speed control of dc motors driven from a phase-controlled rectifier:

1. Because of increased ripple in phase-controlled armature current compared to that which occurs in operation of a motor on pure direct current, higher copper losses for a given torque must be expected.
2. The harmonic content of the phase-controlled armature voltage waveshape produces additional iron losses in the machine.
3. Another factor arises from the back emf of a dc motor load. With a resistive or inductive load, maximum SCR current flows when the conduction angle of each cycle is maximum ($\alpha = 0$). Provided the SCR is capable of handling the current at the maximum conduction angle into a resistive load, it has no problem carrying the lower currents which occur at shorter conduction angles. However, a dc motor may be called on to deliver its maximum torque at very low speeds. This requires the SCR to furnish high average current at a low conduction angle. Because the current rating of the SCR is dependent on conduction angle, this condition of high current during a short conduction angle thus becomes the limiting SCR rating in contrast to the linear resistive type load. A smoothing choke in series with the motor can spread out the conduction angle for a given torque requirement and hence can increase current handling capability of a given SCR.

9.11 Phase-Locked-Loop Control of DC Motors

The phase-locked-loop (PLL) principle used in communications systems has been introduced into motor speed control schemes. Such a system is shown in Figure 9.11. The phase detector provides an output voltage proportional to the difference in phase and frequency of the reference and feedback signals. The high-frequency components of this voltage are filtered out by a low-pass filter. This output voltage is the control voltage for the drive control circuit. When the motor is running at the same speed as set by the reference signal, the shaft speed encoder (ac tach generator) also provides a signal of the same speed as set by the reference signal. Then the two signals will be locked together with a phase difference. Hence the output of the phase detector will be a constant voltage proportional to this phase difference. Any disturbance which changes the speed of the motor will immediately result in a new phase angle position for the feedback signal. The resultant change in phase detector output voltage will vary the output of the motor drive circuit in such a direction as to retain the locking of the reference and feedback signals. It is evident that the motor will always run at a fixed speed irrespective of the load on it, provided the phase difference between the two signals does not exceed 180°. By changing the reference signal frequency or by introducing a divide-by-N counter in the feedback path, the motor can be operated at any desired set speed.

A permanent magnet dc motor or a brushless dc motor (three-phase stator winding with permanent magnet rotor with Hall-effect shaft position detector) may be used as the motor as shown in Figure 9.11.

The PLL control is used in computer disk or tape drives and in any dc position control system.

In the brushless dc motor, the Hall-effect generator is used to sense the shaft position, and a frequency proportional to the motor speed is fed back to the input of the phase detector as shown in Figure 9.11. In the permanent magnet dc motor, an optical shaft encoder could be used.

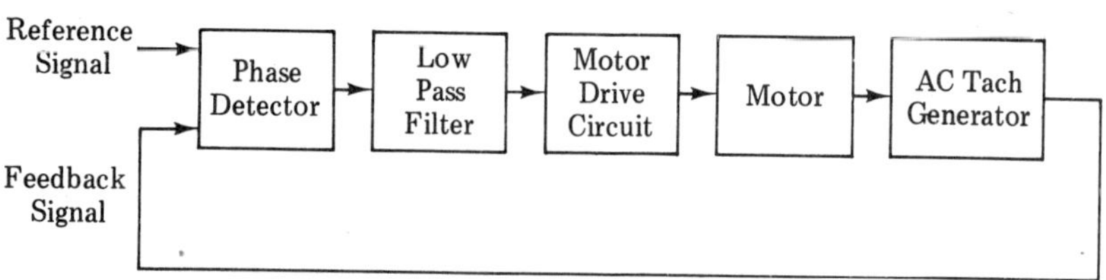

Figure 9.11. PLL motor speed control system.

9.12 Induction Motor Speed Control

For many years attempts have been made to devise satisfactory variable-speed-drive embodying motors having the robustness and simplicity associated with squirrel-cage induction motors. The dc commutator motors are used in the great majority of cases where control of speed is required because they operate with high efficiency over a wide range of conditions, and their speed can be varied in a relatively simple manner by controlling either the field current or armature voltage. In many applications, however, the commutator of the dc motor is disadvantageous; it requires periodic maintenance, which is often inconvenient because access to the commutator is difficult, or because interruptions while in service are undesirable; or, at worst, it cannot be used at all in certain environmental conditions, as, for example, in the case of a pump motor, the rotor of which may actually run in the liquid.

The ac motors, in particular squirrel-cage induction motors, are generally much more satisfactory mechanically than dc motors in many applications, but unfortunately they are basically inflexible as regards speed, since their speed is directly related to their supply frequency, which is normally not variable.

Unlike the dc motor, which is a doubly fed machine in the sense that the stator (field) and the rotor (armature) are both energized by direct current, the squirrel-cage induction motor is singly fed. Only one section, the stator, receives its power directly from the ac mains. The induction motor is essentially a constant-speed energy converter; under normal operating conditions the speed changes only slightly between no load and full load. The speed of an induction motor is definitely "tied to" the line frequency, and to the number of poles for which the machine is wound. Speed variation is fundamentally possible in a squirrel-cage induction motor by employing: (1) a variable-frequency source, (2) varying the voltage applied to the motor, and (3) one of several pole-changing schemes.

9.13 Pole Changing for Speed Control

When a squirrel-cage induction motor is connected to a constant-frequency source, its speed N, in revolutions per minute, is given by

$$N = \frac{120f}{P}(1 - s)$$

where f = frequency of supply

s = slip

P = number of poles

The speed is inversely proportional to the number of poles for which the stator is wound. The foregoing expression for N is the basis for several pole-changing methods of control, which are suited to applications that do not require continuous stepless adjustment and may be driven at two or more base speeds.

9.14 Speed Control by Voltage Variation

The speed of an induction motor may be continuously controlled by controlling the voltage applied to the stator winding. The output torque of an induction motor is proportional to the square of the voltage applied to its stator. This method of speed control changes the output torque of the machine, and hence the speed, but the efficiency falls off drastically with reduction in speed making the method unsuitable for most applications.

9.15 Frequency Changing for Speed Control

An excellent way to control the speed of an induction motor is to vary the frequency of the supply voltage to the stator. With this method it is, of course, necessary to provide a separate source whose frequency and voltage can be adjusted simultaneously and in the correct proportion to each other. The simultaneous variation is desirable because flux densities in the motor which are proportional to volts per cycle, should be kept essentially constant to obtain the proper rated torque. An installation of this type achieved by machines is expensive and cumbersome, since it requires a motor-generator set, which frequently consists of an adjustable-speed dc unit coupled to an alternator or frequency converter. Speed control is, however, virtually stepless over an extremely wide range. The introduction of thyristors has resulted in renewed interest in static power-frequency changers for the speed control of ac motors. The original development of power-frequency changers during the period between 1930 and 1940 made use of grid-controlled mercury arc rectifiers or thyratrons. The operation of the thyristor is similar to that of the grid-controlled mercury arc rectifiers or thyratrons, and it is possible, therefore, to make extensive use of circuits previously developed. The main difference in operation between the two types is that the thyristor is fired by gate current while the mercury arc rectifier or thyratron is controlled by grid voltage.

9.16 Various Schemes of Induction Motor Speed Control

The speed of an induction motor has to be controlled from an ac or dc source. This section gives different schemes for the control of induction motor and are shown in Figure 9.12. In each case the motor is supplied with a variable frequency and variable voltage.

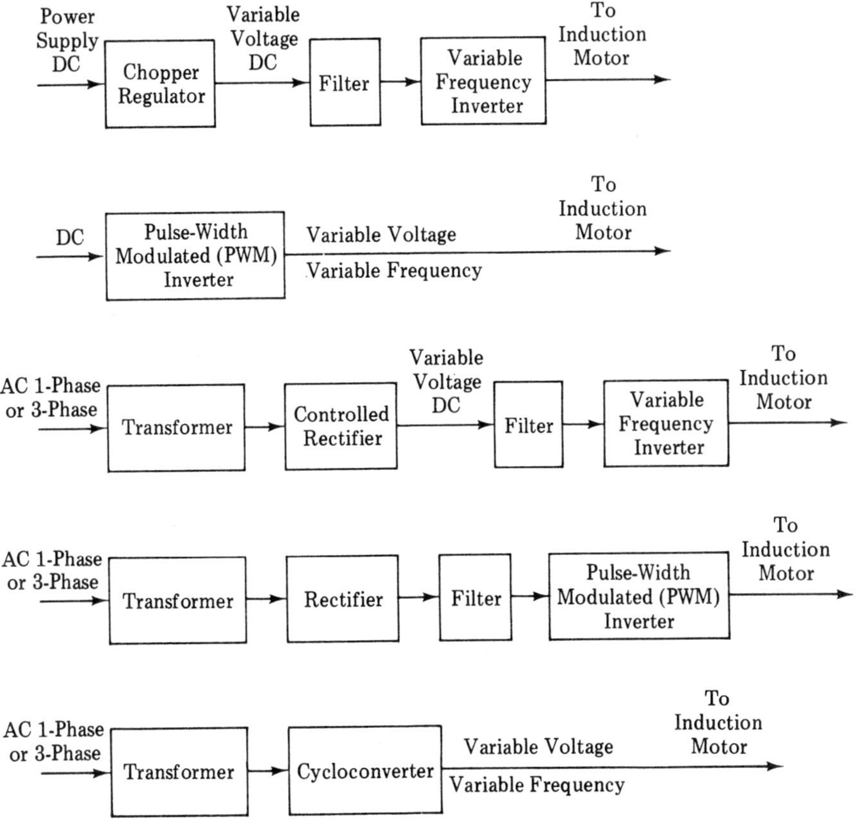

Figure 9.12. Schemes for induction motor speed control.

9.17 Constant Frequency Operation of Induction Motor

The induction motor operating from a constant frequency, constant voltage source is basically a constant speed motor. The equivalent circuit of the induction motor under constant frequency operation is shown in Figure 9.13.

The equivalent circuit parameters are as follows:

V_1 = stator supply voltage per phase

R_1 = stator resistance

X_1 = stator reactance

I_1 = stator current

I_2 = rotor current

I_m = magnetizing current

X_m = magnetizing reactance

R_2 = rotor resistance

X_2 = rotor reactance

$$\frac{R_2(1 - s)}{s} = \text{load as a function of slip}$$

$$S = \text{slip} = \frac{N_s - N_R}{N_s}$$

where N_s = synchronous speed of the motor in rpm

N_R = actual rotor speed in rpm

P_g = air gap power

The torque-speed characteristics of the induction motor is shown in Figure 9.14. At start the motor speed is zero, and the starting torque

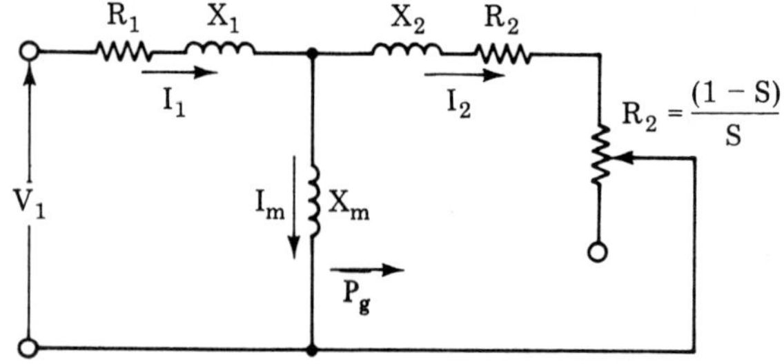

Figure 9.13. Per phase equivalent circuit of a three-phase induction motor.

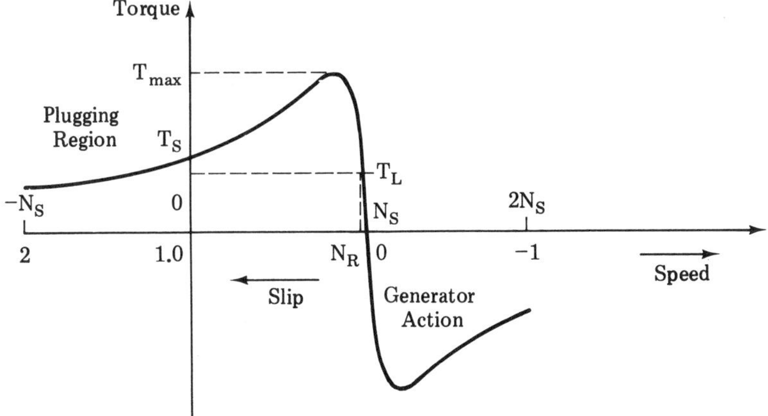

Figure 9.14. Torque-speed characteristic of the three-phase induction motor at constant voltage and frequency.

is shown as T_s. The breakdown torque of the motor is T_{max}. T_L is the rated load torque when the motor is running at a speed of N_R rpm. If the induction motor is driven at a speed greater than N_s (synchronous speed) of the rotating magnetic field, the motor operates with a negative slip and the machine operates as generator.

If the motor is rotating opposite to the direction of rotating magnetic field the slip is greater than unity. The motor operates in this region if the stator field is reversed by changing the phase sequence of the stator supply. This reverses the direction of the rotating magnetic field. Under this condition the machine is quickly brought to a stop, and if the supply is not disconnected, the motor starts to rotate in the opposite direction. This method of braking or rapidly reversing the induction motor is known as plugging.

9.18 Analysis of the Equivalent Circuit Under Steady-State at Constant Frequency

The important characteristics of the induction motor like current, input and output power, losses, speed, starting torque, and maximum torque may be determined from the equivalent circuit.

From the equivalent circuit the total power P_g transferred across the air gap for a three-phase motor is given by

$$P_g = 3I_2^2R_{2/s} \tag{9.25}$$

The total rotor copper loss is

$$3I_2^2R_2 \tag{9.26}$$

The internal mechanical power P developed by the motor is therefore

$$P_{internal} = P_g - \text{rotor copper loss} = (1 - s)P_g \qquad (9.27)$$

The internal electromagnetic torque, $T_{internal}$ is given by $P_{internal}$ divided by angular velocity of the motor, ω_R in radians/sec, or

$$T_{internal} = \frac{P_{internal}}{\omega_R} = \frac{P_{internal}}{\omega_s(1 - s)} \qquad (9.28)$$

where ω_s = synchronous speed in rad/sec

ω_R = mechanical speed of motor in rad/sec

From Equations 9.27 and 9.28,

$$T_{internal} = \frac{P_g}{\omega_s} \qquad (9.29)$$

9.19 Variable Frequency Operation of Induction Motor

The constant frequency equivalent circuit of Figure 9.13 has to be replaced by a variable frequency equivalent circuit as shown in Figure 9.15. As the frequency of the supply to the stator is varied, the stator reactance, rotor reactance, and the magnetizing reactance change as a function of frequency. If the supply voltage is held constant and the frequency is reduced to lower speed, the current in the motor would increase as the impedance of the equivalent circuit is a function of frequency. Hence for variable frequency operation of the induction motor, the voltage and frequency of the supply has to be changed in proper ratio to keep constant current in the motor. The power and torque developed by the motor is a function of the current in the motor.

The equivalent circuit parameters are all referred to the stator side. It is clear from the equivalent circuit that the reactances of the motor vary as a function of frequency.

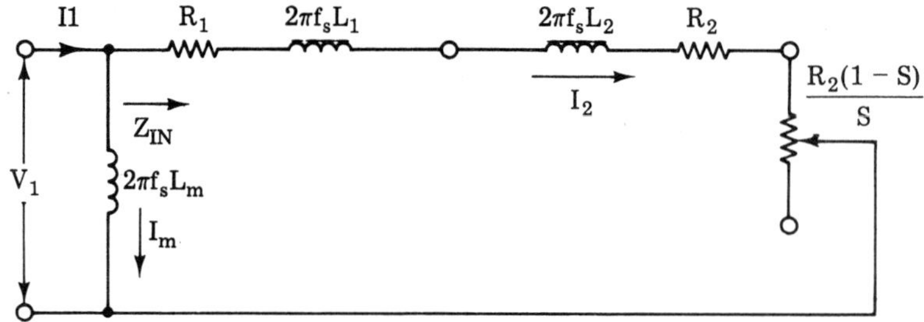

Figure 9.15. Approximate per phase equivalent circuit of a three-phase induction motor.

The total impedance of the equivalent circuit is given by

$$Z_{in} = \left[(R_1 + \frac{R_2}{s}) + j2\pi f_s(L_1 + L_2) \right] \tag{9.30}$$

where f_s = stator frequency in Hz. When the input frequency is high (near 60 Hz), the equivalent circuit impedance is predominantly inductive, and the input impedance varies proportional to frequency. Hence, the supply voltage to the motor has to be varied proportional to frequency to maintain a constant torque. If the input frequency is low (below 40 Hz), the resistance of the motor is not negligible compared to its reactance, hence the impedance of the motor does not decrease proportional to the frequency. Hence, the voltage applied to the motor should not be reduced proportional to the frequency at low frequencies. The volts/Hz must be higher at low frequencies than at higher frequencies.

9.20 Analysis of Induction Motor Operation at Variable Frequency

The induction motor torque-speed characteristics are analyzed at variable frequency using Figure 9.15. The current I_2 flowing in the rotor of the motor is given by

$$I_2 = \frac{V_1}{\sqrt{(R_1 + R_2/s)^2 + 4\pi^2 f_s^2 (L_1 + L_2)^2}} \tag{9.31}$$

The internal power developed by the motor is given by

$$P_{internal} = 3I_2^2 \frac{R_2(1 - s)}{s} \tag{9.32}$$

Neglecting losses in the motor,

$$P_{internal} = P_{output} = P_o \tag{9.33}$$

The motor output torque is given by

$$T_o = \frac{P_o}{\omega_R} = \frac{3}{\omega_R} I_2^2 \frac{R_2(1 - s)}{s} \text{ newton-meter} \tag{9.34}$$

But, $\omega_R = (1 - s)\omega_s$. Therefore,

$$T_o = \frac{3}{\omega_s} \cdot \frac{R_2}{s} \frac{V_1^2}{(R_1 + R_2/s)^2 + 4\pi^2 f_s^2 (L_1 + L_2)^2} \tag{9.35}$$

From Eq. 9.35 the torque T_o is a function of slip for given values of V_1 and ω_s.

The maximum values of torque and the slip at which they occur may be determined by differentiating output torque T_o with respect to s in Eq. 9.35, equating $dT_o/ds = 0$, and solving for s.

The slip at which torque is maximum is given by

$$s = \pm \frac{R_2}{[R_1^2 + (2\pi f_s)^2(L_1 + L_2)^2]^{1/2}} \tag{9.36}$$

Equation 9.35 may be written as

$$T_o = \frac{3}{2\pi f_s} \cdot \frac{P}{2} \cdot \frac{R_2}{s} \cdot \frac{V_1^2}{(R_1 + R_2/s)^2 + 4\pi^2 f_s(L_1 + L_2)^2} \tag{9.37}$$

where P = number of poles. Substituting the positive and negative values of s from Equation 9.36 in the torque expression of Equation 9.37, we obtain the following:

1. Maximum motoring torque,

$T_{m(max)}$

$$= \frac{3}{2\pi f_s} \cdot \frac{P}{4} \frac{V_1^2}{[R_1^2 + 4\pi^2 f_s^2(L_1 + L_2)^2]^{1/2} + R_1} \text{ newton-meter} \tag{9.38}$$

2. Maximum generating torque,

$T_{g(max)}$

$$= \frac{3}{2\pi f_s} \cdot \frac{P}{4} \frac{V_1^2}{[R_1^2 + 4\pi^2 f_s^2(L_1 + L_2)^2]^{1/2} - R_1} \text{ newton-meter} \tag{9.39}$$

At the higher frequency end of operating condition,

$$2\pi f_s(L_1 + L_2) \gg R_1$$

and therefore

$$T_{m(max)} = T_{g(max)} = \frac{3P}{4} \frac{V_1^2}{(2\pi f_s)^2(L_1 + L_2)} \text{ newton-meter} \tag{9.40}$$

Equation 9.39 shows that if volts/Hz (V_1/f_s) is held constant, then the maximum torque or breakdown torque would remain constant.

The starting torque of the induction motor takes place at a slip of one. Substituting s = 1 in Equation 9.35

T_{start}

$$= \frac{3}{2\pi f_s} \cdot \frac{P}{2} \frac{R_2 V_1^2}{(R_1 + R_2)^2 + (2\pi f_s)^2(L_1 + L_2)^2} \text{ newton-meter} \tag{9.41}$$

If the stator supply frequency, f_s, is large

$$2\pi f_s(L_1 + L_2) \gg (R_1 + R_2)$$

Hence, from Equation 9.41

$$T_{start} = \frac{3P}{2} \cdot \frac{V_1^2 R_2}{(2\pi f_s)^3 (L_1 + L_2)} \tag{9.42}$$

In the upper frequency range, if V_1/f_s is held constant, the starting torque would decrease with increasing frequency.

If the stator supply frequency, f_s, is small,

$$2\pi f_s (L_1 + L_2) \ll (R_1 + R_2)$$

Hence, from Equation 9.41

$$T_{start} = \frac{3P}{2} \frac{V_1^2 R_2}{2\pi f_s (R_1 + R_2)^2} \text{ newton-meter} \tag{9.43}$$

In the lower frequency range, if $V1/f_s$ is held constant, the starting torque would increase with increasing frequency.

Hence it is seen from the above discussion, that there is an optimum value of frequency at which the starting torque is maximum.

If the stator supply frequency is increased to run the motor above base speed (or rated speed), the supply voltage cannot be increased above the rated value, and hence the output torque of the motor would fall above rated speed. The torque-speed characteristics of the in-

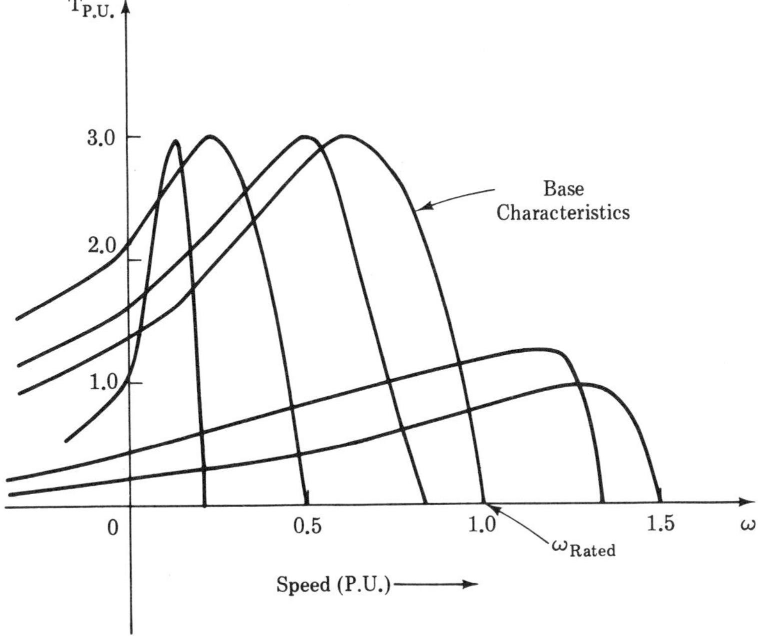

Figure 9.16. Induction motor—variable frequency operation.

duction motor operating from a variable frequency source is shown in Figure 9.16.

9.21 Regenerative Braking of Induction Motor

The induction motor could be slowed or stopped by feeding back power into the ac source. This is referred to as regenerative braking. When the motor is running at a steady speed, if the supply frequency is reduced, the machine would instantaneously run with a negative slip (generating mode) and slow down. The machine would ultimately slow down and run as a motor with reduced speed. The operation of the motor under regenerative braking condition is explained in Figure 9.17. Initially the motor is operating at point a on the curve producing a torque of T_1 (supply frequency of f_{s1} and synchronous speed of ω_{s1} radians/second). As soon as the supply frequency is reduced (f_{s2} and synchronous speed of ω_{s2} radians/second), the motor is operating in the generating mode at point b on the dotted curve. The regenerating torque T_2 slows down the machine and the machine runs as a motor with reduced speed at point c, producing a torque of T_1.

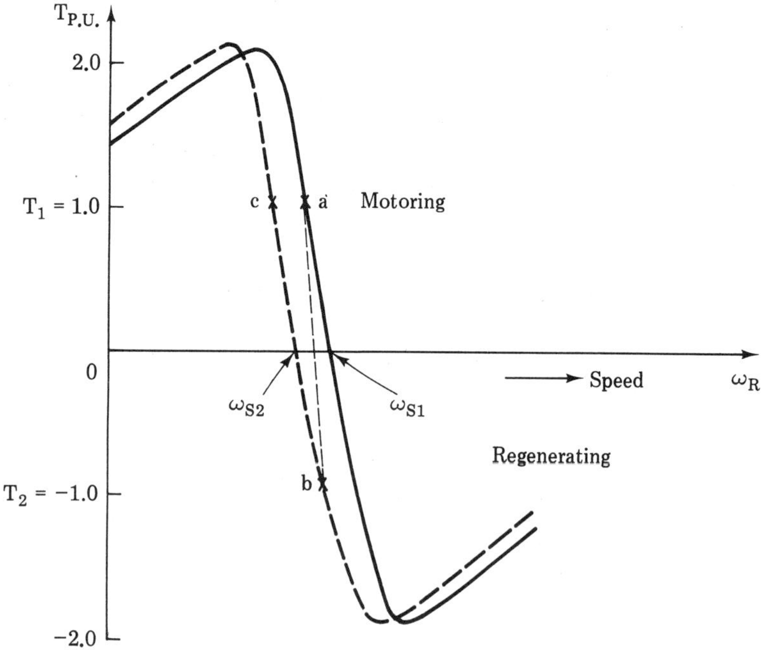

Figure 9.17. Induction motor—regenerative braking.

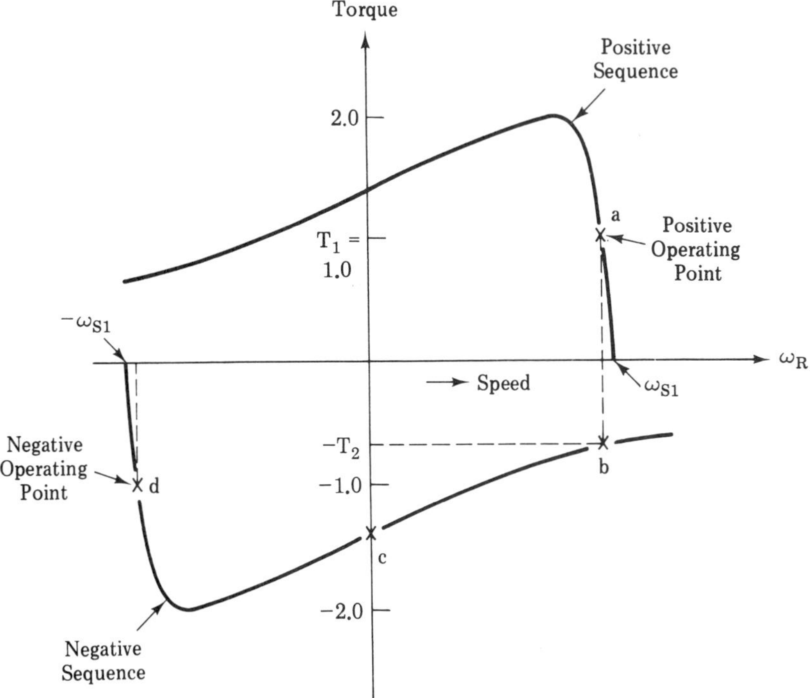

Figure 9.18. Induction motor plugging.

9.22 Induction Motor Plugging

If the phase sequence of the stator supply is reversed, the rotating magnetic field of the stator is reversed. Under this condition the motor is operating in the generating mode and develops a negative torque as is shown in Figure 9.18. The motor is quickly brought to a stop (point c on curve), and if the supply is not disconnected, the motor starts to rotate in the opposite direction and the new operating point is d. This method of rapidly reversing the induction motor is known as plugging.

9.23 Operation of Induction Motor From Nonsinusoidal Voltage Source

The induction motor running from static inverters is subjected to nonsinusoidal voltage waveforms to its stator. The time harmonics present in the applied voltage result in current at the harmonic frequencies.

These harmonic currents increase the losses in the motor. The harmonic currents also are responsible for producing speed and torque ripple in the motor. The output voltage of a three-phase inverter is shown in Figure 8.9 which is supplied to the stator of the induction motor. The expression for the voltage is given by

$$V(t) = \sqrt{2}\,(V_1 \sin \omega t + V_5 \sin 5\omega t + V_7 \sin \omega t + V_k \sin k\omega t) \quad (9.44)$$

The amplitude of the harmonic voltages decreases as order of the harmonics. The important harmonics are the fifth and the seventh.

The performance of the induction motor as explained earlier is obtained from the steady-state equivalent circuit. With nonsinusoidal supply voltage the equivalent circuit has to be modified to represent every harmonic present in the supply voltage. The equivalent circuits for the fundamental, the fifth, and the seventh harmonics are shown in Figure 9.19.

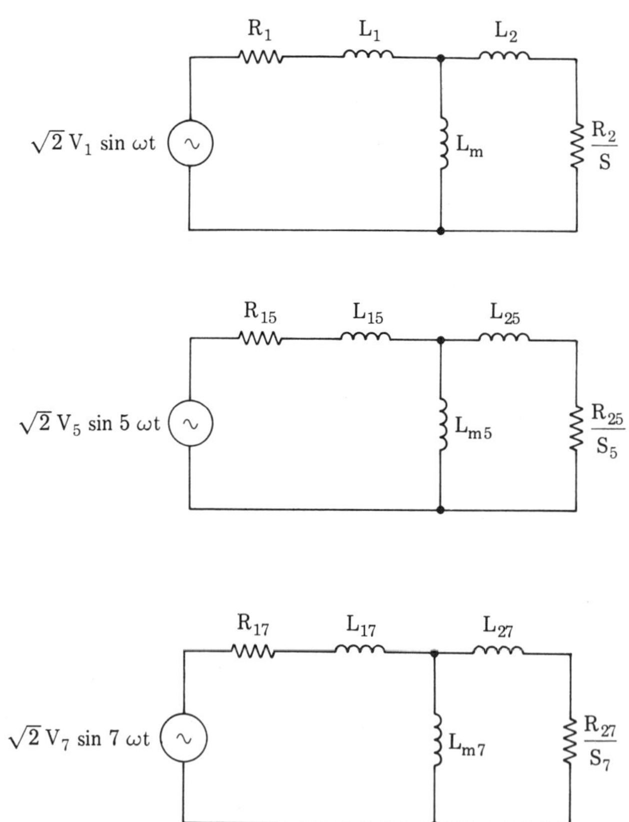

Figure 9.19. Equivalent circuits for induction motor with nonsinusoidal excitation.

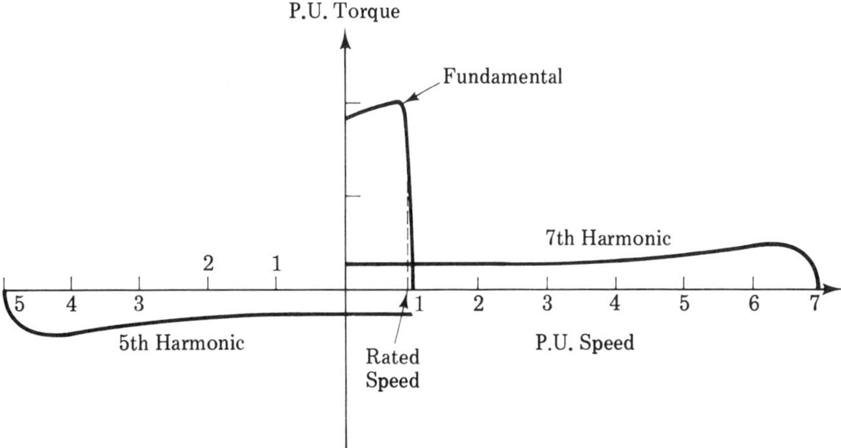

Figure 9.20. Speed-torque curves for the fundamental, 5th and 7th harmonics.

The torque-speed curves of the motor for the fundamental, the fifth harmonic and the seventh harmonic are shown in Figure 9.20. The motor operates at rated torque and rated speed depending on the fundamental frequency. The slip for the fifth harmonic current is given by

$$\frac{5N_s - N_R}{5N_s} = 0.8$$

where N_s = synchronous speed for the fundamental

N_R = mechanical speed of rotor

Similarly the slip for seventh harmonic component is close to unity. Hence for the harmonics it is almost locked rotor condition, when the motor is running at its rated speed. The torque for the fifth harmonic is negative and that of the seventh harmonic is positive. This alternation in sign applies to the subsequent harmonics. Hence the harmonic torques tend to cancel each other.

The interaction of each harmonic rotor current with the air-gap flux of another harmonic will result in pulsating torques. The pulsating torques are generally small and not very significant for most applications.

9.24 Equivalent Circuit Parameters at Variable Frequency

The parameters of the equivalent circuit (Figure 9.15) are not constant at variable frequency. Driving the motor from an inverter introduces

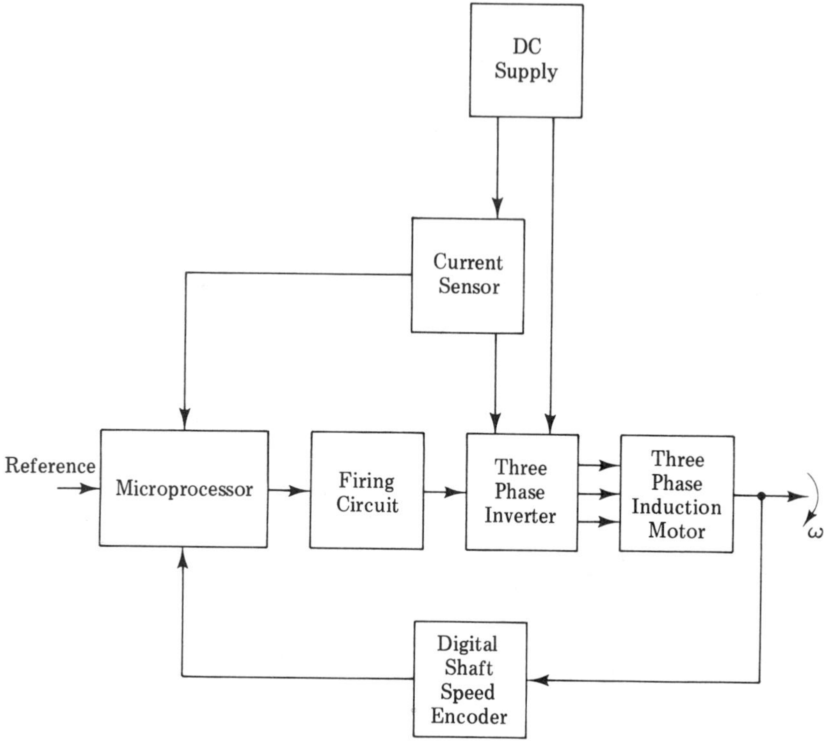

Figure 9.21. Induction motor speed control system.

additional deviations of the equivalent circuit parameters due to harmonics. The rotor circuit resistance and inductance, R_2 and L_2 change as frequency is raised from dc due to the deep-bar effect.

Another factor changing the same parameters is the saturation of the bridge over the rotor slots due to load current. Also harmonic currents increase leakage flux, producing higher saturation in portions of the motor. The magnetizing reactance X_m and leakage inductances (L_1 and L_2) are effected by harmonic currents.

9.25 Microprocessor Based Speed-Control System Using Induction Motor

A closed-loop control of induction motor using digital feedback is shown in Figure 9.21. The digital shaft encoder produces pulses proportional to the speed of induction motor. These pulses control the microprocessor which is controlling the firing circuit of the three-phase inverter.

chapter

TEN

AC-to-AC Converter Cycloconverter

This chapter discusses the operation of a phase-controlled cycloconverter, which is used in the generation of a variable-frequency, variable-voltage supply. The variable-frequency source using the cycloconverter principle may be used for the control of ac motors (induction or synchronous) or as a variable-speed constant-frequency (VSCF) source in the aircraft.

The advent of high-power semiconductor devices for power conversion and control has opened up new fields of application of these devices. As device ratings continue to increase, as dynamic characteristics become even more favorable, and as totally new devices become available, solid-state power conversion technology will continue its phenomenal growth with ever expanding horizons. The introduction of thyristors has resulted in renewed interest in static power-frequency changers. The original development of power-frequency changers during the period between 1930 and 1940 made use of grid-controlled mercury arc rectifiers or thyratrons. The operation of the thyristor is similar to that of the grid-controlled mercury arc rectifier or thyratron, and it is possible, therefore, to make extensive use of circuits previously developed. The main difference in operation between the two types is that the SCR is fired by gate current while the mercury arc rectifier or thyratron is controlled by grid voltage.

10.1 Operating Principle of SCR Cycloconverter

In Figure 10.1 is shown a typical power circuit for a three-phase to three-phase cycloconverter using SCRs for switching. The input to the cycloconverter is from a three-phase, 200V, 60 Hz supply. The cycloconverter as shown in Figure 10.1 consists of 18 SCRs. The function of the cycloconverter is to convert a higher input frequency al-

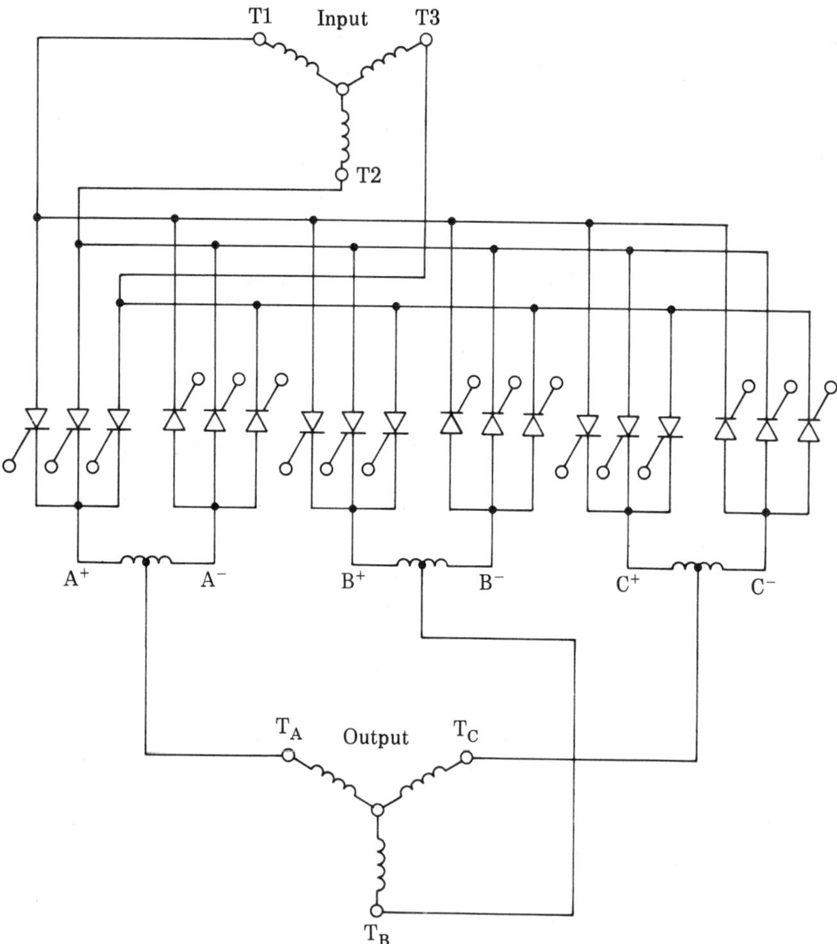

Figure 10.1. Power circuit for typical three-phase to three-phase cycloconverter.

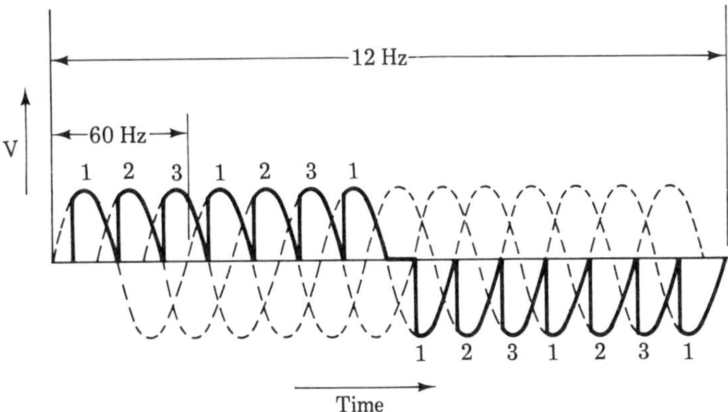

Figure 10.2. Synthesis of waveform for output frequency of 12 Hz from 60 Hz three-phase supply.

ternating current or voltage into a lower output frequency alternating current or voltage. In essence the cycloconverter is a number of switches which are opened and closed at such instants that the output frequency wave is synthesized out of small segments of a number of input frequency waves. This process of opening and closing switches produces a jagged output frequency wave which may be smoothed by a filter if so required.

The cycloconverter illustrated consists of six groups of three-phase half-wave circuits. One half of these groups is called "positive groups," A+, B+, C+, and the other half is called "negative groups," A−, B−, C−. The function of the positive groups is to carry current during the positive half-cycle of the output frequency wave, and the negative groups carry the current during the negative half-cycle of the output-frequency wave. The output frequency is determined by the length of time the positive groups and the negative groups are each allowed to carry current. A synthesis of the waveform is shown in Figure 10.2.

The firing of the SCR is such that, at all times, phase commutation is achieved; i.e., the commutation of current from one SCR to the next is natural due to the voltage difference between the SCRs being of correct polarity. The anode of the SCR next in order of firing is always at a higher potential in the direction of current flow than the SCR which was conducting.

To reduce the harmonic content of the output voltage wave, the firing signals to the SCRs must be controlled in such a way that the output voltage contains a large fundamental component and a small harmonic voltage. This may be accomplished by deriving the firing signals from two sources, that is, the high-frequency voltage of the

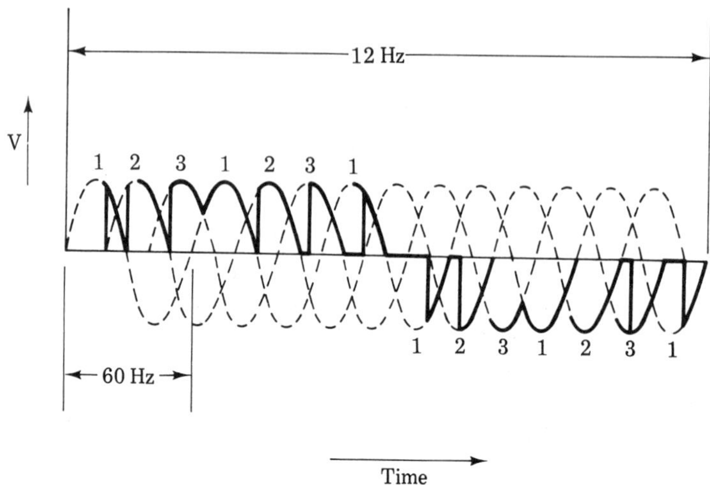

Figure 10.3. Waveform with reduced harmonic content at 12 Hz.

supply and the low-frequency voltage of the reference-wave generator (Figure 10.4). By use of suitable phase relationships, a nearly linear transfer from reference to average output voltage can be obtained as shown in Figure 10.3. The output voltage of the phase-controlled cycloconverter is varied by varying the delay angle of firing of the SCRs in the cycloconverter.

The schematic diagram of the phase-controlled cycloconverter of Figure 10.1 is shown in Figure 10.5. The positive rectifier group (P) permits current flow during positive half of the output waveform and the negative rectifier group (N) permits current flow during negative

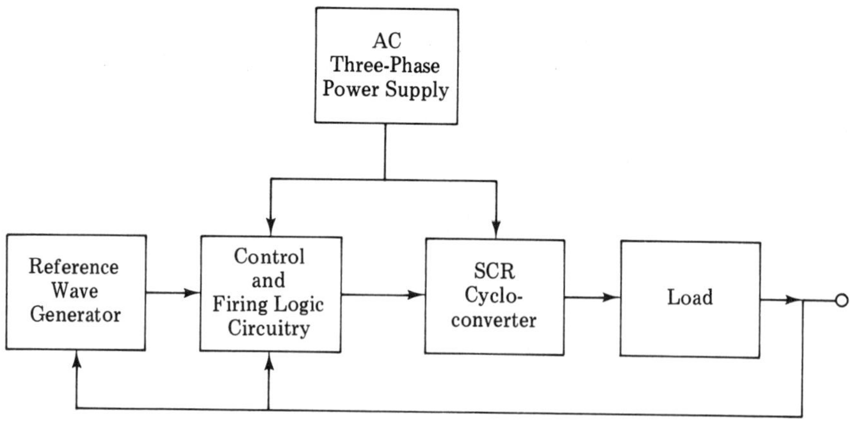

Figure 10.4. Frequency-changer system.

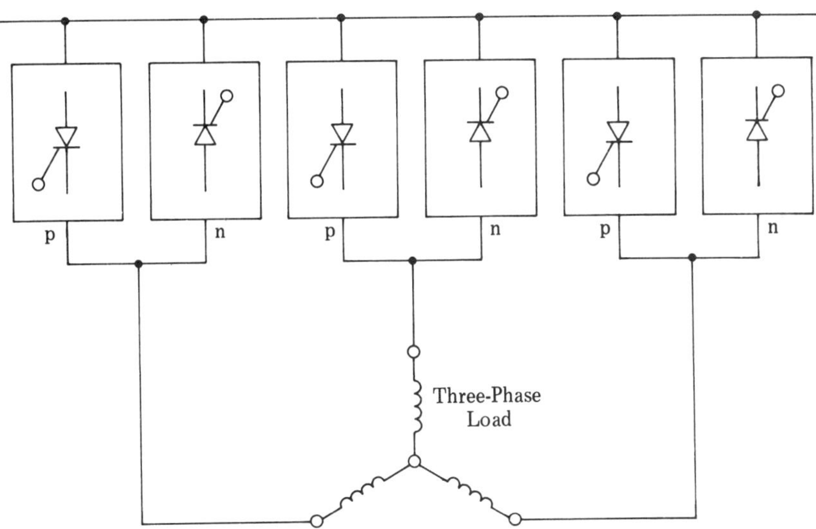

Figure 10.5. Schematic diagram of three-phase to three-phase cycloconverter.

half of the output waveform. Since the positive and negative groups are connected back to back, their average output voltages must always be equal in magnitude and opposite in sign, in order to avoid large circulating currents between groups.

10.2 Circulating Currents in Phase-Controlled Cycloconverter

The cycloconverter delivers low frequency alternating current at the output from a higher frequency input. The circuit could be in a circulating or non-circulating current mode of operation. In the circulating current mode of operation, there is flow of harmonic current between the positive and negative group of rectifiers. In the non-circulating current mode of operation, the flow of current between the two groups are blocked.

10.3 Circulating Current Mode of Operation With Intergroup Reactor

The intergroup reactor as shown in Figure 10.1 is required to limit the harmonic circulating current between the positive and negative groups. The low-frequency output current is limited by reactance x/2, i.e., one-half of the total reactance of the intergroup reactor. If two halves of the reactor are tightly coupled, the harmonic circulating current are limited by 4nx (reactance) where n is the order of the harmonics. The intergroup circulating currents are undesirable, as they increase the losses in the circuit and also increase the current ratings of the thyristors.

10.4 Non-circulating Current Mode of Operation

In a modern cycloconverter circuit, the intergroup circulating currents are suppressed. The suppression of the circulating current is achieved by blocking the gate firing pulses to all the thyristors in the rectifier not delivering load current. If an overload or fault current is detected, then all gate firing pulses are removed to protect the thyristors. The suppression of the circulating current improves the efficiency and displacement factor of the cycloconverter.

10.5 Principle of Operation of Frequency Changer

The block diagram of a frequency-changer system is shown in Figure 10.4. Shown are the primary functional parts of a three-phase to three-phase frequency changer. These include a reference-wave generator, a firing and control logic circuit, and a phase-controlled SCR cycloconverter. The input is derived from an ac three-phase power supply, and the output can supply a three-phase load. The reference-wave generator provides a signal of the desired output frequency (taking into account the load speed) to the firing and control logic circuit. Selection and logical firing of the SCR power switches in the cycloconverter are controlled by firing and control logic circuitry to reproduce the reference frequency at the output. A more detailed discussion of the frequency-changer system is given in the next section.

10.6 Principle of Operation of Three-Phase Static-Frequency Changer System

The principle of operation is discussed with reference to Figure 10.6. The phase-delayed pulse generator generates pulses from each phase of the three-phase ac supply. The phase delay of the generated pulses with respect to the corresponding phase voltage can be continuously varied from 0° to 180° by means of the dc control voltage. A pulse is

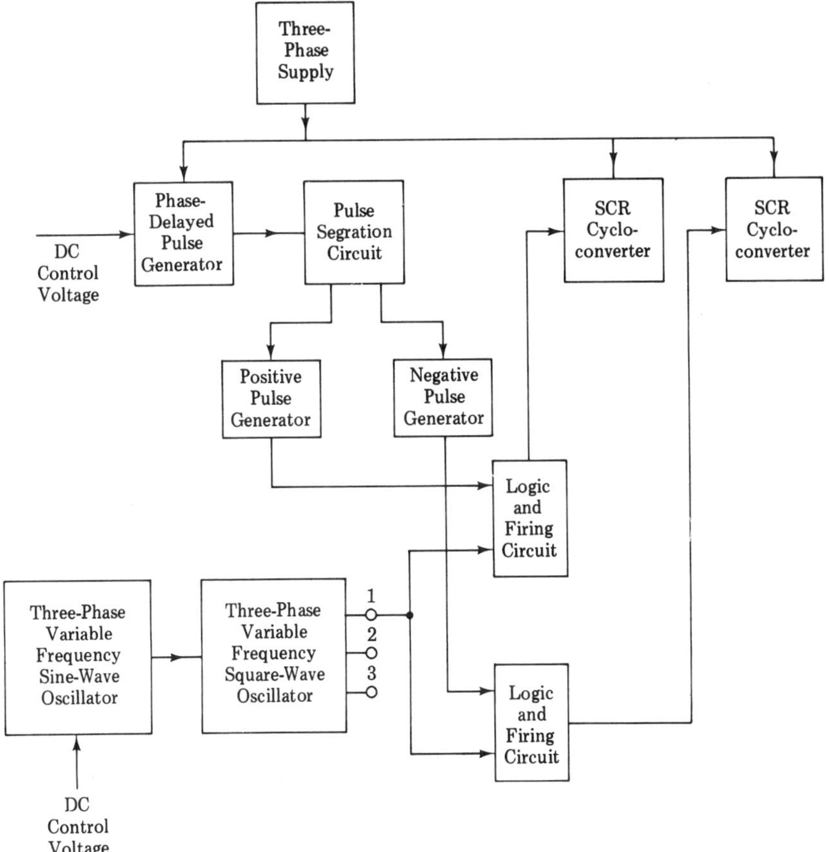

Figure 10.6. Frequency changer (Phase I).

generated on both the positive and negative half-cycles of the ac phase voltage. Thus the pulse repetition period of each generator is 8.3 msec (= 120 pulses/second). The pulses corresponding to the positive half-cycles (hereafter referred to as positive pulses) of Phase I and the positive pulses of Phase II are 120° out of phase. The positive pulses of Phase II and the positive pulses of Phase III are also 120° out of phase. The pulses corresponding to the negative half-cycles of each phase (hereafter referred to as negative pulse) have a similar phase relationship. The pulse segregation circuit of each phase ensures that the positive pulse generator receives pulses generated only from the positive half of the ac supply voltage. Similarly, the negative pulse generator receives pulses from the negative half of the supply voltage. The positive and negative pulses of the pulse segregation circuit are fed to the corresponding logic circuits. The three-phase outputs of the low-frequency oscillator are connected to the corresponding logic circuits. The output frequency of the sinewave oscillator is continuously variable with the dc control voltage. The phase sequence of the oscillator is also dependent on the dc control voltage.

Each logic circuit produces a pulse when pulses of the correct polarity are applied to its two inputs at the same time. The output of each logic circuit triggers its associated blocking oscillator. Each blocking oscillator produces a floating output pulse which is used to fire its associated SCR of the cycloconverter scheme. A floating firing pulse is essential as the gate and cathode of each SCR are not isolated.

Each phase of the cycloconverter consists of six SCRs. The three SCRs in each phase, whose anodes are directly linked to the supply, rectify the positive half-cycle of each phase of the supply. Similarly, the SCRs whose cathodes are directly linked to the supply rectify the negative half-cycle of each phase of the supply. Thus at the point A an alternating voltage is produced which represents Phase I of the frequency-changer output. Similarly, alternating voltages are produced which form Phases II and III of the frequency changer output.

10.7 Gate Control of SCRs in Cycloconverter

In order to control the output voltage of the cycloconverter, it is necessary to control the phase of the SCR firing pulses. Many alternative principles exist for achieving this end. The cosine wave crossing control principle for determining the firing instants can be regarded as being the most natural one for the cycloconverter, as it produces the minimum possible total distortion of the output waveform. Other methods of generating firing pulses as explained in Chapter 11 could also be used.

10.8 Harmonics in Cycloconverter Voltage Waveform

There are two types of cycloconverters used in practice depending on the method of commutation. One method uses natural commutation of thyristors and the second method uses forced commutation principle. The natural commutated thyristor cycloconverter can only produce output frequency, f_o, which is lower than the source frequency, f_s, and the force commutated cycloconverter can also generate output frequency, f_o, higher than source frequency, f_s.

The forced commutated circuit is more complex and the output waveforms are poor. Hence the majority of cycloconverters are naturally commutated.

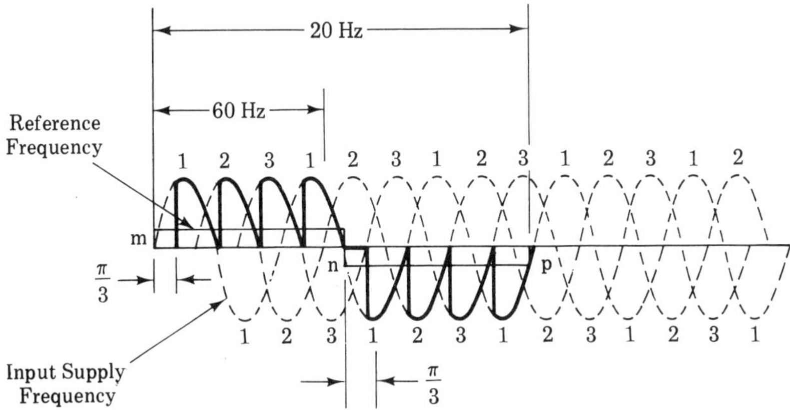

Figure 10.7. Synthesis of waveform for output frequency of 20 Hz from 60 Hz three-phase supply.

The synthesis of a phase-controlled cycloconverter (natural commutation) output waveform at a frequency of 20 Hz from a 60 Hz three-phase supply is shown in Figure 10.7 using square-wave modulation. For simplicity, let us consider the synthesis of Phase I of the output waveform.

The positive half of the low-frequency reference wave is present between m and n. Hence the positive half-cycle of the synthesized low-frequency output waveform is made up from the individual portions of the 60 Hz ac supply. The firing delay associated with each portion is $\pi/3$ in this instance. The firing delay in this particular case is quite arbitrary, but in a control loop would be controlled by the required amplitude of the output wave. The negative half-cycle is similarly made up from the individual portions of the 60 Hz supply between points n and p. The firing delay in this case is also $\pi/3$. Thus, Phase I of the low-frequency output voltage is produced. The positive and negative half-cycles of the low-frequency output wave are identical.

10.9 Open-Loop and Closed-Loop Induction Motor Speed Control

By far the best method of controlling the speed of an induction motor is to change the frequency of the stator supply voltage. This method is limited by the fact that a variable-frequency source of supply is essential. This poses a considerable problem on the design side. This is offset, however, by the very much improved performance obtainable

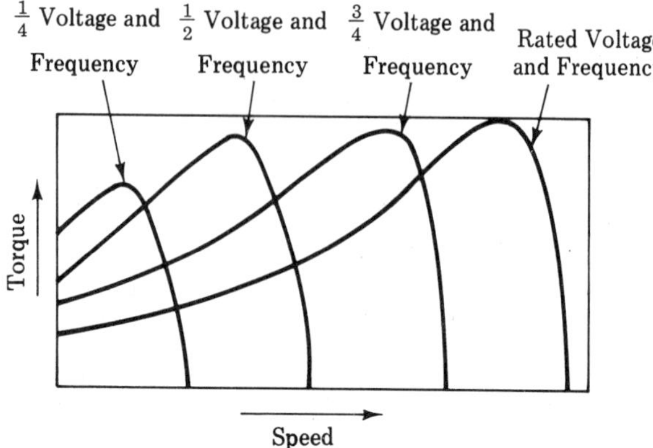

Figure 10.8. Torque-speed characteristics of squirrel-cage induction motor with varying supply frequency.

from such a system. For example, in addition to other advantages, a change in frequency results in a direct change in speed. Figure 10.8 illustrates the torque-speed characteristics obtained for various frequencies, by changing the stator supply voltage in proportion to the frequency. It should be noted that the motor torque falls if the supply voltage is reduced in the same proportion as the supply frequency. Figure 10.9 illustrates the torque-speed characteristics obtained for various frequencies, assuming the stator flux to be held constant by adjusting the supply voltage in a suitable relationship to the frequency. It is thus possible to operate the motor efficiently over a wide range of speeds from zero up to maximum rated speed, and to obtain the full-rated torque over the whole of this range.

The frequency-changer system using the cycloconverter principle

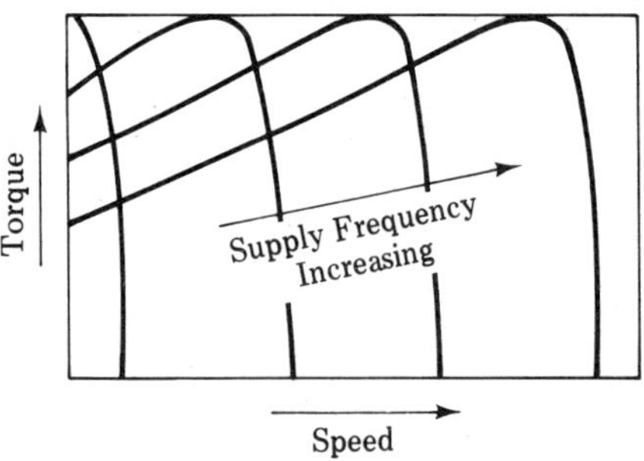

Torque – speed characteristics of squirrel-cage induction motor with varying supply frequency (voltage adjusted to maintain constant torque)

Figure 10.9. Torque-speed characteristics of squirrel-cage induction motor with varying supply frequency.

discussed in this chapter could be used to control the speed of induction motors. An open-loop speed control system is shown in Figure 10.10.

Open-Loop Speed Control

The oscillator frequency can be set to any value between zero and 30 Hz. Hence the output frequency of the frequency-changer could be varied between 0 and 30 Hz. Assume the motor is rated at 220V, 60 Hz, 1750 rpm. If the oscillator output frequency is adjusted to 6 Hz, the voltage supplied to the motor must also be reduced to approximately 220 × (6/60)V in order to obviate saturation. In fact, the voltage supplied is required to be somewhat higher to maintain the rated torque. The amplitude of the output voltage of the cycloconverter is set to the required value by adjusting the delay of the firing pulses with respect to the ac anode voltage.

Closed-Loop Speed Control

The cycloconverter can also be used in a closed-loop manner to control the speed of the induction motor. Figure 10.11 shows the block dia-

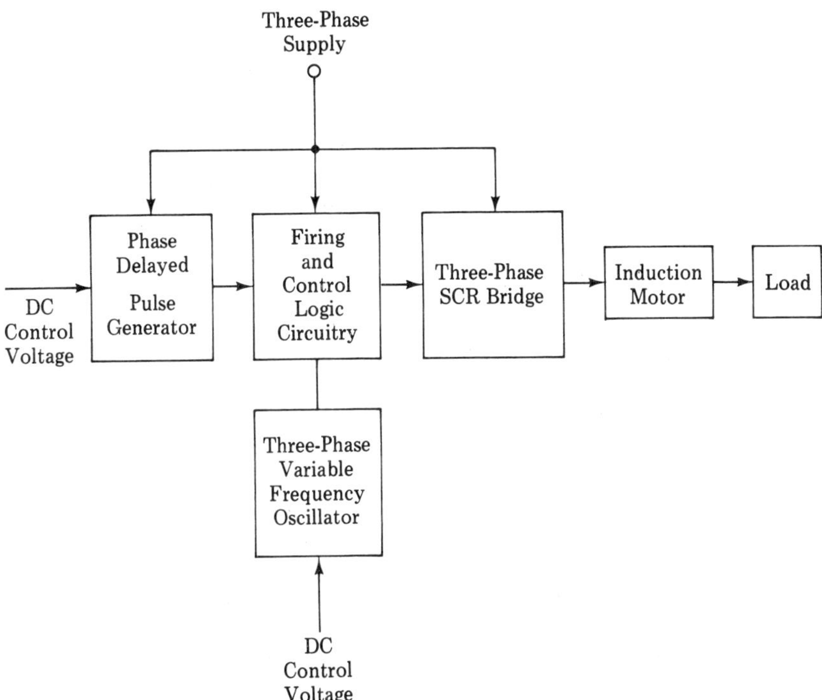

Figure 10.10. Open-loop speed control system.

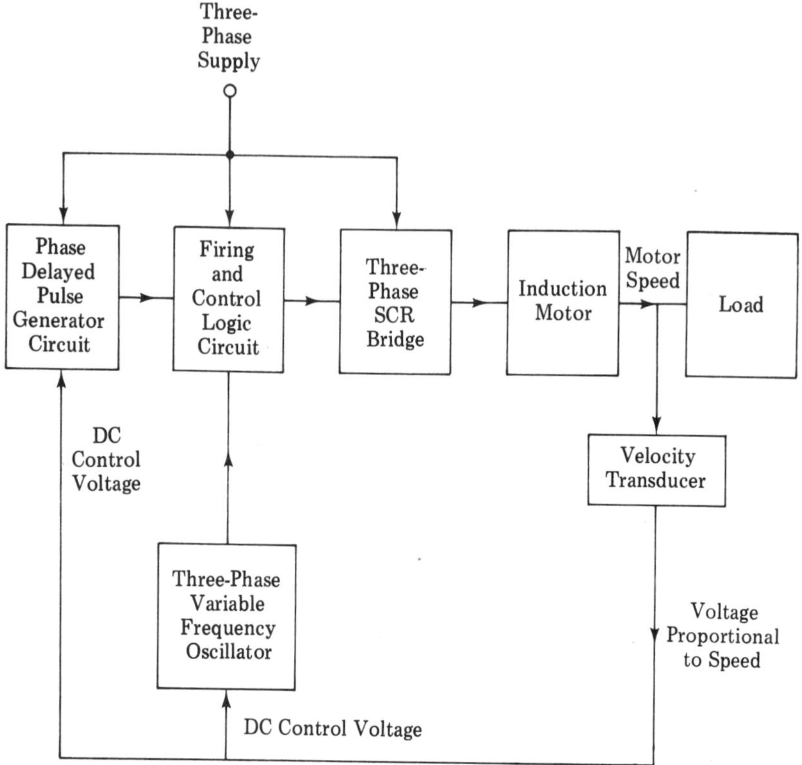

Figure 10.11. Closed-loop speed control of induction motor.

gram of a closed-loop speed control system. The motor speed is sensed by a dc tachogenerator. The required speed is proportional to a dc reference voltage. A reduction in motor speed results in a reduction in the voltage generated by the velocity transducer (tachogenerator). The tachogenerator output voltage is fed back to control (1) the variable-frequency oscillator after comparison with the reference input-voltage, and (2) the phase-delayed pulse generator, i.e., V1 and V2, respectively. The two potentiometers permit appropriate scaling of these two voltages before they are fed back to the respective control circuits. The dc voltage fed back to the variable-frequency oscillator is proportional to motor speed, and the difference between this voltage and the dc reference input voltage provides the error which together with the dc reference controls the frequency. The voltage amplitude of the cycloconverter output must also be controlled proportional to motor speed; this is achieved by comparing V1 with a dc bias voltage, and using the difference to control the delay of the phase-delayed pulse generator unit.

The closed-loop control scheme thus automatically compensates for any load fluctuations, or required cycloconverter output frequency and amplitude for any given speed.

10.10 Output Voltage of Three-Phase Controlled Rectifier

Due to artificial delay in phase commutation, represented by the firing delay angle α, the rectifier phase whose voltage is e_1 is forced to carry the load current during the period represented by the time interval between the points R and S, instead of during the natural period of current conduction represented by the time interval between the corresponding points P and Q (Figure 10.12). As the result of this retarded commutation, the output voltage of the rectifier system is given by the mean value of the phase voltage e_1 during the period from R to S, that is, by the mean ordinate of the shaded area under the curve. The individual phase voltages are each expressed by $e = \sqrt{2}E \cos \theta$, so that the mean output voltage of the rectifier system is given by

$$V_d = \frac{3}{2\pi} \int_{\alpha - \pi/3}^{\alpha + \pi/3} \sqrt{2}E \cos \theta \, d\theta = \sqrt{2}E \cdot \frac{3}{\pi} \sin \pi/3 \cos \alpha$$

$$= 1.18E \cos \alpha = V_{do} \cos \alpha \quad (10.1)$$

where $V_{do} = 1.18E$ is the maximum value of the output voltage when $\alpha = 0$.

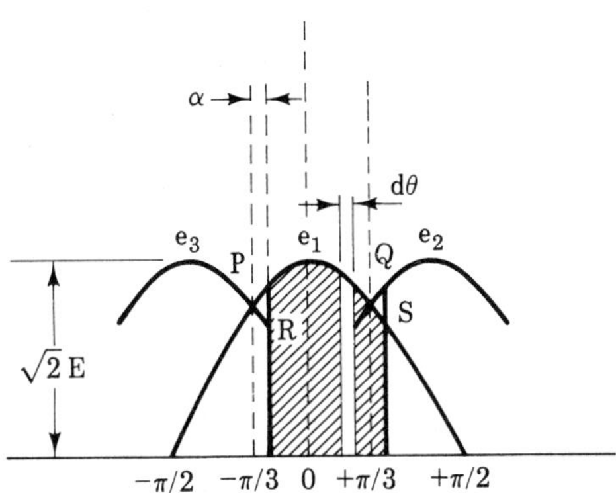

Figure 10.12. Waveforms for three-phase rectifier with firing delay angle α.

10.11 Output Voltage of Cycloconverter

In the case of the cycloconverter the peak value of the lower frequency output voltage is equal to the mean value of the ac output voltage of the equivalent rectifier system. Considering a normal rectifier circuit the average output voltage V_d is, as shown in the previous section, given by

$$V_d = \sqrt{2}E \cdot \frac{3}{\pi} \sin \pi/3 \cos \alpha \tag{10.2}$$

where E is the rms value of the input voltage and α is the firing delay angle of the rectifier past the point of natural commutation.

With the previously mentioned assumption, V_d represents the peak output voltage of the cycloconverter, therefore

$$V_o\sqrt{2} = \sqrt{2} \cdot \frac{3}{\pi} \sin \pi/3 \cdot E \cos \alpha$$

or

$$V_o = \frac{3}{\pi} \sin \pi/3E \cos \alpha = .82E \cos \alpha \tag{10.3}$$

where V_o is the rms value of the cycloconverter output voltage. Hence maximum rms output voltage, V_{max}, is given by

$$V_{max} = 0.82E \tag{10.4}$$

where $\alpha = 0$.

If r is the voltage reduction factor, that is, the ratio of the actual amplitude to the maximum amplitude, then the rms output voltage of the cycloconverter becomes

$$V_o = r \frac{3}{\pi} \sin \frac{\pi}{3} E \tag{10.5}$$

The lower frequency voltage can thus be varied from zero to its maximum value by varying the firing delay angle α.

10.12 Variable-Speed Constant-Frequency Supply

A variable-speed constant-frequency (VSCF) source is needed in some applications, where it is required to produce an accurately regulated fixed frequency power from a variable frequency power source. In this type of application, an SCR cycloconverter could be ideally used for frequency conversion. In aircraft power conversion this method is used. The basic principle of operation is shown in Figure 10.13.

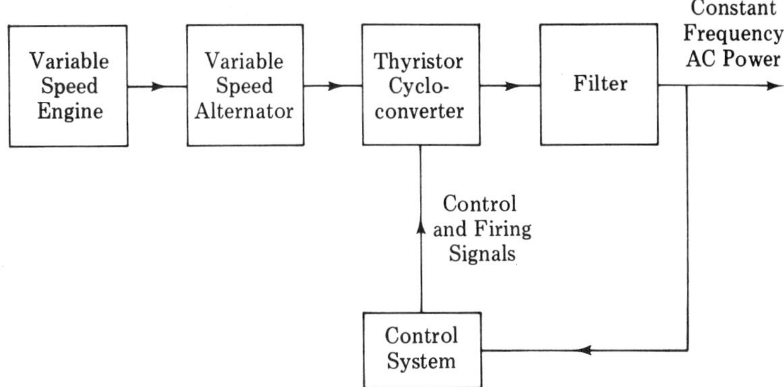

Figure 10.13. Block diagram of a variable-speed constant-frequency (VSCF) system using thyristor cycloconverter.

The variable-speed jet engine drives a high-speed alternator which generates a variable-frequency power output. The general practice has been to use a hydraulic constant speed coupling device between the engine and the alternator, thereby driving the alternator at a constant speed to generate a constant frequency. A more modern approach using solid-state cycloconverter principle is shown in Figure 10.13.

Here the variable speed alternator driven by the variable speed jet engine produces a variable-frequency output. The output frequency is quite high, as the alternator is driven at a high speed by the high-speed jet engine. This variable-frequency power is converted to a constant-frequency power, by means of an SCR cycloconverter and an electric filter.

In aircraft application, weight and reliability are of prime importance. The main advantages of the solid-state VSCF system are that it requires less maintenance and its electrical performance is superior to that of a hydraulic system. If the low frequency oscillator controlling the cycloconverter is designed to be stable, then the output frequency of the VSCF system is absolutely stable.

chapter

ELEVEN

Thyristor Trigger Circuits

A steep wavefront or a fast rising current pulse is required for best triggering of thyristors (SCRs). Before a circuit used to trigger an SCR can be designed in detail, the SCR gate specifications must be considered.

The proper design of firing circuits of an SCR involves three important considerations: 1) choice of a suitable circuit which will supply the firing signal under the required circuit conditions, 2) determination of the maximum voltages and currents which the firing circuit can supply to ensure that the maximum allowable gate ratings are not exceeded, and 3) determination of the minimum voltages and currents which it will supply to the gate to ensure that the SCRs will be reliably fired under all conditions. Less trigger current is required at higher temperatures while a higher value is required at lower temperatures. It has also been shown that trigger current requirements decrease slightly as the anode to cathode voltage is increased. Whenever precise timing or phase control is required, it is desirable, due to the temperature and anode effects on the control characteristics and also due to production spreads, to fire with trigger current of steep wavefront. If the gate signal is a slow variation of a dc voltage, or a sinewave, the firing point of the SCR will vary because of changes in the junction temperature. If accurate firing is required, the slowly changing signal must be converted into a relatively fast pulse at suitable signal level, this pulse being then used to trigger the device. Also the trigger signal must be derived from a low-impedance circuit as it is required to supply appreciable current for firing, especially for large thyristors. If the pulse duration is shorter than six to eight microseconds, a larger amplitude is required to accomplish firing. There is approximately an inverse correlation between pulse length and amplitude, over the range of 0.2 μsec to 2 μsec pulse widths. There seems to be a practical limit in the vicinity of 0.1 μsec pulse width, under which firing is not possible.

11.1 Turn-On Effects

Unpredictable effects during switching of silicon-controlled rectifiers have been noticed by designers of high-frequency inverters and pulse modulators, where high values of di/dt occur. Such effects are due to a turn-on phenomenon: the junctions are turned-on, at first, only over a small area, near the gate lead, and the process then spreads until the whole junction has been turned on. Hence the initial current density is high.

When the SCR is turned on the anode to cathode voltage does not drop immediately; instead, the voltage appears to drop exponentially. This effect is more prominent with high values of di/dt.

The high initial current density and high voltage drop generate considerable power within a small area, thus raising the temperature of the junctions. If the temperature exceeds 125°C there is a temporary falling off of the characteristics; this is especially noticeable in forward breakover voltage, turn-off time, and dv/dt. In inverters, this effect appears as unexpected triggering. If the heat becomes excessive, it can permanently damage the thyristor characteristics. Such failure shows as a small hole in the pellet from anode to cathode near the gate lead.

The high local values of current density can be satisfactorily decreased by placing a saturable reactor in series with the thyristor; this limits the current for several microseconds to a low value, until the reactor saturates. During the delay period the current density is reasonably low when the bulk current flows.

11.2 Gate Characteristics of SCR

SCR gate specifications must be considered for the design of trigger circuit. The important gate characteristics are as follows:

1. The device has to be protected from damage such as peak forward gate voltage V_g, peak forward gate current, I_g, peak instantaneous gate power, V_gI_g, and maximum average gate power $(V_gI_g)_{avg}$.
2. Specifications giving the minimum gate currents and voltages required to trigger the SCR. These values are temperature dependent.

The trigger circuit design is further complicated by the fact that there is a significant variation of the gate V-I characteristics and the minimum gate voltages and currents required to trigger the SCR for individual devices of the same type.

A typical gate characteristic of an SCR is shown in Figure 11.1.

The thyristor gate drive area for reliable triggering must be bounded by ABCD as shown in Figure 11.1. If the gate current I_g and the gate voltage V_g intersect within the boundaries of this curve, the device will be successfully triggered. The gate current requirement increases as the temperature decreases. The shaded area shows locus of all trigger points. I_{gmin} for various temperatures are shown.

The maximum gate voltage allowed is shown by line B. The maximum instantaneous gate power dissipation is shown by curve C. The

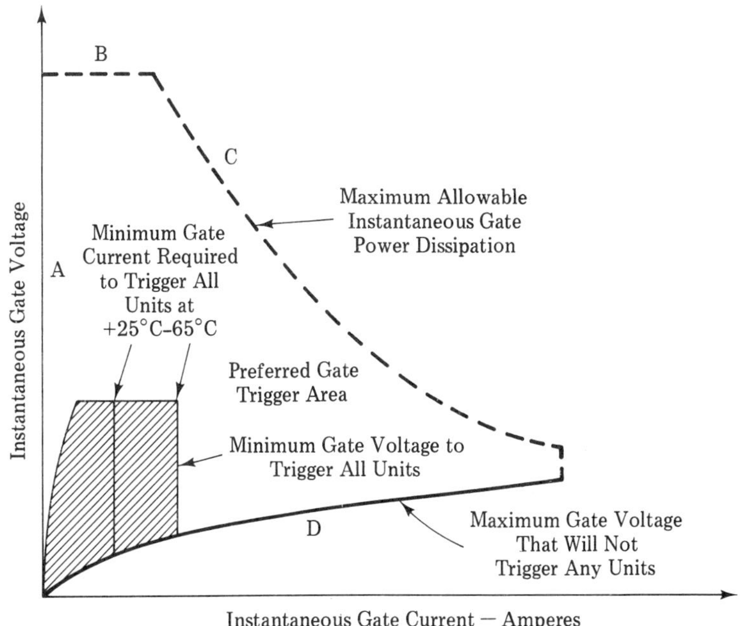

Figure 11.1. Gate trigger characteristics of an SCR.

maximum gate voltage that will not trigger the device is shown by curve D. The gate of a thyristor is often triggered by a pulse or a train of pulses to keep the gate power dissipation low.

Typical trigger circuits using UJT and PUT are shown in Chapter 3.

Bibliography

1. P. Sen, *Thyristor D.C. Drives,* John Wiley and Sons, 1981.
2. S. B. Dewan and A. Straughen, *Power Semiconductor Circuits,* John Wiley and Sons, 1975.
3. D. F. Geiger, *Phaselock Loops for D.C. Motor Speed Control,* John Wiley and Sons, 1981.
4. F. M. Gardner, *Phaselock Techniques,* John Wiley and Sons, 1979.
5. J. M. D. Murphy, *Thyristor Control of A.C. Motors,* Pergamon Press, 1973.
6. B. R. Pelly, *Thyristor Phase-Controlled Converters and Cycloconverters,* John Wiley and Sons, 1971.
7. R. S. Ramshaw, *Power Electronics,* Chapman and Hall, London, 1973.
8. B. D. Bedford and R. G. Hoft, *Principles of Inverter Circuits,* John Wiley and Sons, 1964.
9. F. F. Mazda, *Thyristor Control,* John Wiley and Sons, 1973.
10. A. Kusko, *Solid-State D.C. Motor Drives,* M.I.T. Press, 1969.
11. E. W. Kimbark, *Direct Current Transmission* Vol. I, John Wiley and Sons, 1971.

12. F. E. Gentry, F. W. Gutzwiller, etc., *Semiconductor Controlled Rectifiers*, Prentice-Hall, 1964.
13. J. Seymour, *Semiconductor Devices in Power Engineering*, Issac Pitman, London, 1968.
14. R. M. Davis, *Power Diode and Thyristor Circuits*, Cambridge University Press, England, 1971.
15. T. Takeuchi, *Theory of SCR Circuit and Application to Motor Control*, Tokyo, 1968.
16. J. W. Motto, Jr., Editor, "Introduction to Solid State Power Electronics", Westinghouse Electric Corporation, 1977.
17. J. Schaefer, *Rectifier Circuits, Theory and Design*, John Wiley and Sons, 1965.
18. W. McMurray, *Theory and Design of Cycloconverters*, M.I.T. Press, 1972.
19. J. D. Harnden and F. B. Golden, *Power Semiconductor Applications Vols. I and II*, IEEE Press, 1972.
20. SCR Manual 6th Edition, General Electric Company, 1979.
21. SCR Designers Handbook, Westinghouse Electric Corporation, 1970.
22. C. Adamson and N. G. Hingorani, *High Voltage Direct Current Power Transmission*, Garraway, London, 1960.
23. B. J. Cory, Editor, *High Voltage Direct Current Converters and Systems*, MacDonald, London, 1965.

INDEX